ESL VIDEO LIBRARY

FOCUS ON INNOVATORS AND INNOVATIONS

PATRICIA DUFFY

REGENTS/PRENTICE HALL
Englewood Cliffs, New Jersey 07632

Acquisitions Editor: **Nancy L. Leonhardt**
Development Editor: **Stephanie Karras**
Editorial Assistance: **D. Andrew Gitzy, Terry TenBarge**
Production Editor: **Janice Sivertsen**
Prepress Buyer: **Ray Keating**
Manufacturing Buyer: **Lori Bulwin**
Cover Supervisor: **Marianne Frasco**
Cover Design: **Ruta Kysilewskyj**
Cover Photograph: **© 1992 Arthur Tilley/FPG International**
Interior Design: **Tom Nery**
Technical Support: **Molly Pike Riccardi, Freddy Flake**

Printed in the United States of America
10 9 8 7 6 5 4 3 2

ISBN 0-13-007105-6

Prentice-Hall International (UK) Limited, *London*
Prentice-Hall of Australia Pty. Limited, *Sydney*
Prentice-Hall Canada Inc., *Toronto*
Prentice-Hall Hispanoamericana, S.A., *Mexico*
Prentice-Hall of India Private Limited, *New Delhi*
Prentice-Hall of Japan, Inc., *Tokyo*
Simon & Schuster Asia Pte, Ltd., *Singapore*
Editora Prentice-Hall do Brasil, Ltda., *Rio de Janeiro*

TABLE OF CONTENTS

DEDICATION

To my brother, Jackie

ACKNOWLEDGEMENTS

The influence and energy of many people helped make the writing of this book possible, and I wish to thank some of them now.

First, I'd like to thank my colleagues in the English department of the United Nation's Language Training Programme for their interest, encouragement, and feedback on this project. Thank you to Elizabeth Kahn, Laura Layton, Cary Vaughan, Bob Burbidge, and Linda Haley.

I'd also like to thank my colleague, Kirit Makita, for his always kind and willing technical assistance, which made it possible for me to fully test my materials with students in the Language Training Programme.

Another thank you to Tivia Weisbrot and Rosalind Stevenson for their crucial computer assistance.

A very big thank you to my colleagues at New York University—to Allan Kent Dart, my original ESL writing mentor, to Maxine Steinhaus, my ever faithful friend and supporter, and to Yatsen Chan, who first inspired me to incorporate authentic video materials into language teaching.

I must certainly thank all of my students in both the United Nation's Language Training Programme and New York University's American Language Institute; it was, of course, their reactions, comments, criticism, and enthusiasm which helped this book take shape.

Finally, for their constant support and understanding, I'd like to pay tribute to my family members: Estelle Duffy, Marie, Christina, and Sean Duffy, Luna Tarlo, and to the memory of my father, John Duffy—himself, a true innovator.

Most of all, I want to say thank you to my husband, Josh Cohen, for his unending, unswerving, and often unbelievable patience and assistance throughout all the ups and downs of this project—no matter where on the globe he happened to be.

INTRODUCTION

The *ABC News* ESL Video Library, a first-of-its-kind series, is the product of an evolving pedagogy in the field of language teaching. Increasingly, language teachers have emphasized the importance of basing teaching materials on authentic sources. *Focus on Innovators and Innovations*, one of the five books in the series, does just that; the book's twelve chapters are based on twelve authentic news broadcasts from internationally-known *ABC News* programs, such as *Nightline, 20/20, Prime Time Live*, and *World News Tonight*.

Educators have long been aware that television news programs contain rich pedagogical resources that teachers and students can "mine" for new vocabulary, idioms, grammatical structures, listening and pronunciation practice. Each segment (chapter) of *Focus on Innovators and Innovations* contains a wide variety of exercises emphasizing all of those diverse skills. Exercises also aid students in exploring, analyzing, discussing, and writing about the content of each of the news broadcasts.

To facilitate the convenient use of this video-textbook package, time codes have been included in the upper right-hand corner of the screen. These time codes are cross-referenced in the text so that teachers and students can easily locate particular program segments and their corresponding textbook exercises.

The text utilizes an interactive approach, providing opportunities for students to work in pairs or small groups and encouraging them to exchange opinions and express ideas from their own cultural perspectives. Teachers and students need not feel compelled to do all exercises in all categories. The assortment of exercises is intended to provide a broad spectrum of choices; students and teachers may choose those most appropriate to their individual or program needs.

In addition to its variety of exercises, *Focus on Innovators and Innovations* contains a variety of subject matter. While the book's overall theme concerns those people and ideas that have been helping to reshape our world, the four units within the text focus on innovators and innovations in four different fields: education, science and technology, social reform, and the arts. The topics in the book have been student tested for interest level and capacity to promote a high degree of verbal interaction in class.

Focus on Innovators and Innovations chapters are divided into five sections: Previewing, Global Viewing, Intensive Viewing, Language Focus, and Postviewing. The purpose of each section is as follows:

Previewing offers students a chance to predict what the segment will cover. This section also helps set the students' schema and enables the teacher to know what knowledge they already possess about the topic. This section also presents essential vocabulary that the students will need to know in order to understand the segment.

Global Viewing develops students' understanding of the key ideas presented in the segment. In this section, students' overall comprehension and skill in determining main ideas are stressed.

Intensive Viewing zeroes in on specific items for comprehension to prepare students for the segment's final tasks. Students practice looking for details in notetaking and cloze-type exercises.

Language Focus gives practice in the vocabulary and expressions that the students will need to use when completing the final task at the end of the segment.

Postviewing provides additional materials, such as related readings and also ideas for projects, which enable students to pull together all the language and content skills they have practiced in the segment.

An additional feature of the videos is that all of the programs (with the exception of the *20/20* program about Andy Warhol) are closed-captioned. If teachers or students have access to a closed-captioned decoder or television set with a decoder chip, it is possible to see the program's dialogue and narration written out on the bottom of the screen.

Meaningful content has been increasingly recognized as a major ingredient in the language teaching/learning recipe. By working through the selection of news programs and corresponding segments in *Focus on Innovators and Innovations,* students may expand their knowledge of a changing world, as they expand their knowledge of a new language. Fortified with such knowledge, learners will be better prepared to interact and perhaps to make their own innovations within the emerging global community.

FOCUS ON INNOVATORS AND INNOVATIONS

Focus on Innovators and Innovations consists of twelve news reports. The overall goal of the text is: (1) to aid in comprehension of the video segment; (2) to encourage critical thinking; and (3) to stimulate the use of the language presented on the video.

In the **Previewing**, **Global Viewing**, **Intensive Viewing**, and **Postviewing** sections, the content of the video generates a variety of viewing activities, such as: predicting, getting the general idea, and notetaking. It is also used as a basis for exercises focusing on vocabulary and grammatical structures such as: the present perfect tense, multiple word verbs, and the article.

The **Postviewing** section includes a wide variety of tasks that allows students to discuss, share their reactions, and gather information related to the particular segment. The following is a summary of the various postviewing activities presented in each segment of the text.

Unit 1: Education

Segment & Title	Functions in Postviewing Tasks
1 Conflict Management	• Agreeing and disagreeing with opinions about conflict resolution • Reading and gathering information about UNESCO • Reacting to a quotation • Writing a reaction • Creating and recording a radio program
2 It Pays to Get Smart	• Agreeing and disagreeing with opinions about education • Reading and gathering information about corporations • Reacting to a quotation • Interviewing a business person

Unit 1: Education (continued)

Segment & Title	Functions in Postviewing Tasks
3 Wild About Learning	• Agreeing and disagreeing with opinions about early childhood education • Reading and gathering information about content-based learning • Forming a committee to write an education program • Interviewing a teacher • Keeping a journal

Unit 2: Science and Technology

Segment & Title	Functions in Postviewing Tasks
4 The Food of the Future	• Agreeing and disagreeing with opinions about new food technology • Reading and gathering information about food • Reacting to a quotation • Writing a reaction
5 Virtual Reality	• Agreeing and disagreeing with opinions about virtual reality • Reading and gathering information about a science fiction writer • Creating an original application of virtual reality • Reacting to a quotation • Creating and recording a radio program
6 Conquering Fear at Camp Courage	• Agreeing and disagreeing with opinions about helping chronically ill children • Reading and gathering information about heart transplants • Reacting to a quotation • Researching programs that help handicapped children

Unit 3: Social Reform

Segment & Title	Functions in Postviewing Tasks
7 The Power of Mimi	• Agreeing and disagreeing with opinions about reforming criminals • Reading and gathering information about the prison system • Reacting to a quotation • Creating and recording a radio program
8 Annie DeMartino	• Agreeing and disagreeing with opinions about the homeless • Reading and gathering information about the homeless • Reacting to a quotation • Writing a letter • Researching the problem of homelessness
9 James Grant	• Agreeing and disagreeing with opinions about children in developing countries • Reading and gathering information about UNICEF • Reacting to a quotation

Unit 4: The Arts

Segment & Title	Functions in Postviewing Tasks
10 Andy Warhol	• Agreeing and disagreeing with opinions about Warhol's art • Reading and gathering information about other artists • Reacting to a quotation • Creating and recording a radio program

Unit 4: The Arts

Segment & Title	Functions in Postviewing Tasks
11 Martha Graham	• Agreeing and disagreeing with opinions about Martha Graham's work • Reading and gathering information about other dancers • Reacting to a quotation • Comparing two forms of dance • Researching pioneers in modern dance
12 Black Filmmakers	• Agreeing and disagreeing with opinions about stereotypes • Reading and gathering information about Spike Lee • Defining and discussing stereotypes • Reacting to quotation • Writing about minority groups

Options for Self-Study

Although *Focus on Innovators and Innovations* is meant to be used in a classroom setting with a teacher, it is also possible to use this video course for self-study. English-language learners can choose the exercises and activities they feel they need the most, and check their answers in the Instructor's Manual.

Students who have a decoder should make use of the closed-captions by reading the words on the screen. This should enhance comprehension, particularly when a student is working alone. If a decoder is not available, students should refer to the transcripts in the Instructor's Manual.

Segment 1

Conflict Management

from *World News Tonight,* 2/18/90
Runtime: 4:51
Begin: 00:00:39

Some elementary schools have begun teaching school children techniques for resolving disagreements peacefully.

Previewing

KEY QUESTIONS

1. Is it possible for people to live their lives totally free of conflict with other people?
2. Is it possible for people to learn to solve their day-to-day conflicts without arguing or fighting?
3. Can people be educated to live in peace?

DISCUSSION

1. Do you think that fighting between children in elementary school is a problem? Why or why not?

____

____

2. Do you remember fighting or arguing with other children when you were in elementary school? What happened? How were the conflicts resolved?

____

____

3. Do you think schools should teach young children how to solve their problems peacefully? Or should schools confine themselves to teaching academic subjects?

____

____

PREDICTION

Based on the title of the segment, *Conflict Management*, the discussion you've had with your classmates, and your own knowledge, what information do you think will be included in the video segment?

1. ____
2. ____
3. ____
4. ____

ESSENTIAL WORDS TO KNOW

The following sentences are from the video segment. Read the sentences and write your own definition of the *italicized* word(s).

1. In a peaceful schoolyard in Brooklyn, a fight starts to *break out.*

   break out: ____

2. Unlike settlements *imposed* by principals, these resolutions stay resolved because kids confront the real issues.

   imposed: ____

3. The *mediators* have been taught special techniques to help their fellow students solve their own problems.

   mediators: ____

4. Conflict *resolution* is proving to kids that they can control violence in the world and in themselves.

   resolution: ____

5. The mediators ask *disputants* to analyze the problem and find a solution both sides feel is right.

   disputants: ____

Global Viewing

CONFIRMING YOUR PREDICTIONS

00:00:39-00:05:30

Watch the entire video segment. As you watch, put a check (✓) next to any of your predictions that were mentioned.

Prediction 1 ______

Prediction 2 ______

Prediction 3 ______

Prediction 4 ______

MAJOR POINTS/MAIN IDEA

00:00:39-00:05:30

With a partner, watch the entire segment, then write down a few of the major points made in the program.

Compare the major points that you have identified with those chosen by other students in the class. Ask one of the students to go to the front of the room and write on the chalkboard all of the different major points chosen by the class. With the rest of the class, decide which is the most important point. Do any of them sum up the major idea of the program? Write it below in a complete sentence.

THINK, WRITE, AND DISCUSS

00:00:39-00:05:30

Watch the entire segment and answer the questions on page 4. If you wish, you may take notes on a separate sheet of paper. After you've finished, discuss your answers with the class.

1. Is the conflict resolution program working effectively in the schools? Why?

____

____

2. Are teachers or other authority figures involved in resolving students' conflicts? How?

____

____

3. What is the goal of conflict resolution?

____

____

4. Is the subject of conflict resolution taught only at the elementary school level? If not, where else is it being taught?

____

____

Intensive Viewing

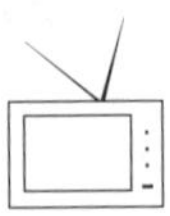

00:00:39-
00:01:15

LISTENING CLOZE

Watch this part of the segment and fill in the missing words. Then compare your answers with those of another student.

Peter Jennings: We have put ______________ ________________ on the American Agenda tonight. Not conflict nor, for that matter resolution in the Persian Gulf, but conflict resolution in the school ________________. We've done so because ______________ interferes with ______________, and it is possible to teach young people how to ______________ without fighting and how to ________________ themselves without the involvement of ________. The news is that schools across the country are training students in the same __________________ used by diplomats, and it is apparently ____________.

THE PROCESS OF CONFLICT RESOLUTION

Watch the part of the segment that talks about how conflict resolution works and fill in the information.

00:02:09-00:02:56

1. The three rules of conflict resolution
   a. The first rule: ______________________________
   b. The second rule: ______________________________
   c. The third rule: ______________________________
2. Techniques students have been taught
   a. ______________________________
   b. ______________________________
   c. ______________________________

In your opinion, are the three rules of conflict resolution and the techniques the students have been taught enough to achieve conflict resolution? Would you add any other rules to this list?

____

____

____

HOW DO THE STUDENTS FEEL?

In this part of the program segment, the students are asked how they felt after becoming mediators. Watch that portion of the program again. What are some of the answers they give? As you listen to the students' answers, take notes below about what they say. Then compare your notes with those of other students in the class.

00:02:57-00:03:20

1. The first student: ______________________________
   ____
2. The second student: ______________________________
   ____
3. The third student: ______________________________
   ____

00:04:12-
00:04:28

TEACHING CONFLICT RESOLUTION

Which universities offer courses in conflict resolution? Which university offers a Ph.D. degree in the field of conflict resolution? Where are these universities located? Write their names and locations here.

	NAME	LOCATION
Universities offering courses in conflict resolution:	______________	not given
	______________	not given
University offering Ph.D. in conflict resolution:	______________	______________

Language Focus

VOCABULARY IN CONTEXT

Sometimes it is possible to guess the meaning of an unfamiliar word by looking at the other words that surround it and guessing the meaning from its context. Guess the meaning of the *italicized* words in the sentences below. First circle the words in the sentence that helped you to guess the meaning. Then write a definition in the space provided. The first one has been started for you.

1. Kids are still (fighting) (verbally;) they're *arguing*.

   arguing: ____

2. It has made a more *serene* environment. The students really believe that we care about them and that we want them to solve their problems.

   serene: ____

3. *Conflict* resolution is proving to kids that they can control violence in the world and in themselves.

   conflict: ____

4. "Okay, then you pushed her once?"
   "I just—I gave her a little *nudge* and said, 'Stay off me.'"

   nudge: ____

5. They've been taught special *techniques* to help their fellow students solve their own problems, such as getting the disputants to describe their feelings.

   techniques: ____

STRUCTURE PRACTICE: ACTIVE/PASSIVE

When changing a passive sentence to an active sentence, the positions of subject and object are reversed. In some cases, however, the object is understood and not written. The object becomes the subject, and we must state the subject in order to write an active sentence. To write the example below as an active sentence, we must identify who taught them ("*They*").

> *They've been taught* special techniques to help their fellow students solve their own problems.
>
> *Teachers have taught* them special techniques to help their fellow students solve their own problems.

The following sentences from the segment are written in passive form. Rewrite them as active sentences, reversing the position of subject and object.

1. They can be mediated by fellow students.

   ____

2. Conflict resolution is taught at a growing number of universities and colleges like Harvard and Stanford.

   ____

3 Students are trained in the same techniques used by diplomats.

   ____

4. Is it conflict resolution if one party is forced to the [negotiating] table by physical force?

   ____

Postviewing

DO YOU AGREE OR DISAGREE?

Work with a small group of students and discuss whether you agree or disagree with the following statements. Be prepared to justify your answers. Choose one member of the group to be the leader. After the small group discussions have finished, the group leaders can share the opinions of the group members with the rest of the class.

1. It's a good idea to teach conflict resolution techniques to children in elementary school.
2. Conflict resolution techniques probably work better with children than with adults.

3. Teaching conflict resolution in elementary school is a waste of time; children need to spend more time learning academic subjects such as mathematics and reading.
4. If all children learned conflict resolution techniques, the world would become a more peaceful place.
5. Places like New York City, which have a lot of potential for social conflict, need conflict resolution courses in their schools; but schools in many other parts of the world don't need to teach students such techniques.
6. Children should not have the role of mediating disagreements between other children; only adults have the maturity to take on this role.

RELATED READING 1

Read the article *Peace Research* to answer the questions that follow.

Peace Research

The relatively recent academic discipline of peace research was originally developed in response to the Cold War conflict. During the Cold War, which started in the 1950s, the nations of the world were dangerously polarized into two camps, East and West. Much of the early peace research was done in Scandinavia and the United States. The International Peace Research Institute was founded in Oslo, Norway, and the Center for Research and Conflict Resolution at the University of Michigan, U.S.A. According to John Galtung of the International Peace Research Institute, "peace research can be conceived as a quest for knowledge pertinent to the preservation of peace" and as "a systematic attempt to understand the conditions of peace."

The goal of peace research is "the development of methods of conflict control and conflict resolution, and social integration." Certain of the methods developed as a result of peace research are now being studied at institutions of higher learning around the world, and incorporated into the curriculums of some elementary schools.

QUESTIONS

Write short answers to the questions below. Then underline the part of the article that justifies your answer to the question, and write the number of the question next to it.

1. Why was the academic discipline of peace research developed?

____

____

2. What happened during the Cold War of the 1950s?

3. Where was a lot of the early peace research carried out?

4. Who is John Galtung?

5. How does John Galtung define peace research?

RELATED READING 2

Read the article *UNESCO* to prepare for the task on page 10.

UNESCO

The United Nations Educational, Scientific, and Cultural Organization, or UNESCO, was created in 1946 as an autonomous agency of the United Nations. Its main headquarters is in Paris, France. The aim of UNESCO is to promote world peace by facilitating international collaboration on educational, scientific, and cultural projects. Because those at UNESCO believe that wars start in the minds of human beings, the organization contributes to educational programs that foster intercultural and interracial understanding. The programs also foster relations between people of different nations in the belief that the more connected people feel, the more motivated they will be to resolve disagreements that may arise between them.

UNESCO is also active in promoting world literacy and training in basic skills to aid the development of nations. Throughout the world, UNESCO advises on school curriculum reform, production of textbooks, and educational material. It seeks to incorporate the studies of human rights and peace as essential parts of educational programs.

QUESTIONS

Write three or four questions about the reading on page 9. Begin each of your questions with the words below. Then ask your partner to answer the questions without looking at the text.

1. When ____?

2. Why ____?

3. What ____?

4. Where ____?

To obtain more information about the activities of UNESCO, you may write to the address below. You may also want to ask UNESCO to send you the latest edition of the *Directory of Peace Research and Training Institutions* and the *UNESCO Yearbook on Peace and Conflict Studies*.

UNESCO
7, Place de Fontenoy
75007, Paris, France

WRITE ABOUT IT

One of the students says that learning to be a mediator reminds her of the saying "Before you can change the world, you've got to change yourself." Do you agree or disagree with this statement? Write your reactions. Then share your reaction with your classmates.

PROJECT

Idea # 1

You may wish to learn more about the field of conflict resolution or peace studies. The program mentions that the latter is an area of academic study at such universities as George Mason, Harvard, and Stanford. More and more universities and educational organizations throughout the world are focusing attention on peace studies. Write a letter to one or more of the organizations on page 11 requesting information on this developing field. In your letter, introduce yourself and explain why you feel the area of peace studies is an important one. Be sure to use at least four or five of the vocabulary words that you learned in the course of watching the segment. If you receive an answer to your letter, share the information you receive with your classmates and teacher.

Educators for Social Responsibility
475 Riverside Drive, 9th floor
New York, NY 10115

Japanese Association of Language Teachers (JALT)
"Global Issues in Language Education"
Tottari University
Koyama Tottori City, Japan 680

Washington Teachers of English to Speakers of Other Languages (WAESOL)
Committee on Sociopolitical Concerns
2121 N. 143 Street
Seattle, Washington 98133-7211

United Nations University for Peace
P.O. Box 199
Escazu, Costa Rica

McMasters University Peace Studies Center
DC 229
1280 Main Street West
Hamilton L8S4K1
Ontario, Canada

Idea #2

Work with a small group of students and make a "radio program" composed of your reactions to the video segment *Conflict Management*. Each of you can prepare a statement to comment on a different aspect of the segment. Decide in which order to put the comments. Take time to rehearse your script, correcting sentences as necessary. You may want to ask your teacher to check your written script before you put it on tape to make your "radio program." Be sure to use in your tapescript some of the words and phrases you learned from watching the program. After you complete your program, play it for the rest of the class. If you wish, you may have a class contest to choose the best program.

SUGGESTED READINGS

Grimshaw, Allen, editor, *Conflict Talk*. Cambridge University Press, 1990.

Smoke, Richard and Harman, Willis. *Paths to Peace: Exploring the Feasibility of Sustainable Peace*. Westview Press, 1987.

Segment 2

It Pays to Get Smart

from *20/20,* 9/13/91
Runtime: 15:39
Begin: 00:05:38

Two businessmen decided they could show America how to run schools successfully by applying techniques of business management.

KEY QUESTIONS

1. What is a "good school"? What are the necessary elements of a good learning environment?
2. What can businesses contribute to educational systems?
3. How can educational systems overcome the problems children experience in urban environments?

DISCUSSION

1. What are the criteria for a good learning environment? What countries in the world do you suppose have the best schools? What makes these schools the best?

____

____

2. Some people say that the U.S. educational system is not doing a good enough job of educating its diverse population of students, particularly those who are economically disadvantaged. In your opinion, what are some of the most important features of a good educational system?

____

____

3. What things did you like about the schools you attended? What things could have been changed or improved?

____

____

PREDICTION

Based on the title of the segment, *It Pays to Get Smart*, the discussion you've had with your classmates, and your own knowledge, what information do you think will be included in the video segment?

1. ____
2. ____
3. ____
4. ____

ESSENTIAL WORDS TO KNOW

The *italicized* words will be used in the video. Read the words as they are used in the following sentences and try to come up with your own definitions.

1. They created their own school as an *alternative* to the failing public schools.

   alternative: ____

2. Corporations are in the business of making money, of looking after *profits*.

   profits: ____

3. In the new school, leaders seek *accountability* at every level; if you don't do a good job, you're gone.

   accountability: ____

4. In Chicago, they created their own *inner-city* school.

   inner-city: ____

5. This new school is a kind of *research and development laboratory* where new ideas about teaching and learning are tried out.

   research and development laboratory: ____

Global Viewing

CONFIRMING YOUR PREDICTIONS

00:05:38-00:21:17

Watch the entire video segment. As you watch, put a check (✓) next to any of your predictions that were mentioned.

Prediction 1 _____

Prediction 2 _____

Prediction 3 _____

Prediction 4 _____

MAJOR POINTS/MAIN IDEA

00:05:38-00:21:17

Work with a partner and write down a few of the major points made in the program.

____

____

____

____

____

____

Compare the major points you have written with those written by other students. (One student can list on the chalkboard all of the different points chosen by the class). With the rest of the class, decide which point is the most important, or which sums up the major idea of the program. Write it below in a complete sentence.

____

____

THINK, WRITE, AND DISCUSS

00:05:38-00:21:17

As you watch the program, think about the questions on page 16. Take notes to answer them whenever possible. After the program, discuss your answers with the other members of the class.

1. Why are Joe Kellman and Vernon Loucks, two businessmen, interested in running a school?

   ____

   ____

2. What do Joe Kellman and Vernon Loucks mean when they say "the future of corporate America is tied to the inner-city"?

   ____

   ____

3. What does Joe Kellman say about the amount of money that has been spent on education? Does he feel the education system could be improved by increasing the amount of money spent on it?

   ____

   ____

4. What effect has attending the Corporate School had on many of the students?

   ____

   ____

5. What is the Corporate School trying to demonstrate to other schools?

   ____

   ____

Intensive Viewing

LISTENING CLOZE

00:07:26-
00:08:08

Watch this part of the segment and fill in the missing words. Then compare your answers with others.

Stone Phillips: Even in places like the Lawndale section of Chicago, one of the worst __________ ________ neighborhoods in the country, _______________ still have dreams, but from here the high-rise world of _____________ and opportunity is a distant, hazy view, at best, of a world that's too often ____________________ to them. Well, recently, two of Chicago's top ____________________ from that distant world decided that the ____________ of corporate America is ________ to the dreams of inner-city children and they could no longer ____________ to be indifferent. They believed good ___________________ was the answer, so they came here, to the heart of Lawndale, to try a new ___________________ venture.

THE BEGINNING OF THE CORPORATE SCHOOL

Watch this part of the segment and fill in the notes. Then compare your notes with those of another student.

00:08:48-00:10:47

1. Kind of company Vernon Louks works for: ________________________
2. Kind of company Joe Kellman owns: ____________________________
3. Their idea for the Corporate School
   a. privately run
   b. ____
4. Enrollment in the school
   a. How many children would attend: ____________________________
   b. How they would be selected: lottery
   c. requirements to attend school: _______________________________
5. How Louks and Kellman convinced corporations to contribute to the school: ____

   ____
6. Companies that contributed to the school:
   a. Sears Roebuck c. __________________________
   b. __________________________ d. __________________________

THE THREE MAIN PRINCIPLES OF BUSINESS MANAGEMENT

In this part of the segment, Stone Phillips says that Joe Kellman and Vernon Louks applied three rules of business management to make the Corporate School run successfully. In the space below, write the three rules and describe how they were applied to the Corporate School.

00:11:20-00:19:25

1. The first rule (Hint: Joe Kellman says, "If you don't fit you're gone.")

   ____

   How it's applied: ____

   ____

2. The second rule (Hint: Joe Kellman says, "Leave them alone and don't handcuff them.")

____

How it's applied: ____________________________________

____

3. The third rule (Hint: Stone Phillips says, "It isn't enough just to teach reading, writing, and arithmetic.")

____

How it's applied: ____________________________________

____

Language Focus

USING THE ARTICLE

The articles *a* and *an* precede the name of one's profession. Articles also precede the names of other roles people take on in life. At the beginning of the segment, you heard some of the students use articles in this way.

"I want to be *a* nurse."
"I want to be *a* lawyer."

Look at the chart explaining when to use "a" or "an."

ARTICLE	WHEN TO USE	EXAMPLE
a	use before words beginning with a consonant sound	a mother, a father, a doctor, a tennis player
an	use before words beginning with a vowel sound	an aunt, an uncle, an actor, an English teacher

Make sentences using the appropriate article (*a, an*) to describe all the roles you play in life at the present time. After you finish, share this information about your partner with the rest of the class. Do the different members of the class have many of the same goals?

Example: *I am a student of English. I am also a son, and a cousin. I am a baseball player.*

Now work with a partner and ask him or her what he or she would like to be in the future. Following the above model, write two (or more) sentences describing what your partner would like to be.

Example: *In the future, he would like to be a doctor, a husband, a father, a good amateur photographer, and a good speaker of English.*

MORE STRUCTURE PRACTICE: *USED TO*

In the program segment, we heard the following lines of dialogue.
"Patrick *used to* have a problem fighting."
"He *used to* be in the office for fighting. He's doing much better now."
Write sentences about bad habits you used to have, but don't have anymore. For example:

I used to bite my nails.
I don't do that anymore.

USED TO VS *BE USED TO*

Compare these two sentences. Do they have the same meaning?

1. They *used to have* a long summer vacation.
2. They *are used to having* a long summer vacation.

What is the difference in the meaning?

In sentence 1, *used to have* means that during a past period, it was their habit to have a long summer vacation.

In sentence 2, *used to having* means that in general, they are accustomed to having a long summer vacation.

Notice the difference in the form of the two sentences. In sentence 1, the base form of the main verb follows *to*. In sentence 2, the verb *be* precedes *used to* and is followed by the gerund form of the main verb.

When you first began your English class, perhaps there were some activities you were not accustomed to doing; maybe by now you feel more comfortable doing these activities because you are used to doing them. Write some sentences about activities you are used to doing now.

Example: I am used to speaking English in class now.

1. ____________________

2. ____________________

3. ____________________

Postviewing

DO YOU AGREE OR DISAGREE?

Work with a small group of students. Discuss whether you agree or disagree with the following statements and why. Choose one student to be the group leader. After you have finished your discussion, the group leader can report the opinions of the group members to the rest of the class.

1. Joe Kellman is right—schools should be run according to corporate management principles.
2. If employees have too much job security, they won't work hard.
3. The Corporate School will have limited success if the children who attend it have a difficult family life.
4. Only the brightest children should go to the Corporate School.
5. The business community is interested in schools like this one for its own reasons; business leaders simply want to manipulate people to fit into the corporate world.

RELATED READING 1

Read the article *Ben and Jerry's Homemade: Good Ice Cream, Good Causes* to answer the questions on page 22.

Ben & Jerry's Homemade:
Good Ice Cream, Good Causes

It is hard to believe that Ben & Jerry's Homemade Ice Cream Company, now a powerful multimillion dollar corporation, began as a small, simple ice cream shop in a little town in Vermont. In 1980 Ben & Jerry's Homemade was started by two idealistic and clever young men named Ben Cohen and Jerry Greenberg, who decided to put their ingenuity and principles to work.

Ben and Jerry believed that a business should do more than just pursue profits. They believed that a business has a responsibility to channel a part of its profits back into the communities it serves. As Ben & Jerry's Homemade has stated in its annual report to shareholders, "Our company believes that business has a responsibility to work actively at improving the quality of life for those working within the business, for the local community, and for the world at large." In keeping with its philosophy, Ben & Jerry's has developed a variety of ways to help improve the quality of life on both national and international levels. One way it has done this is by creating an organization called 1% for Peace. 1% for Peace seeks to redirect 1 percent of the U.S. defense budget (approximately 3 billion U.S. dollars) to educational programs that build understanding among nations by promoting greater intercultural awareness. Ben & Jerry's contributes 7 1/2 percent of its own annual gross income to the 1% for Peace organization. In addition, it has a program for small businesses around America to join this effort by donating 1 percent of their own annual profits to 1% for Peace.

Ben & Jerry's also sponsors programs that provide employment to disadvantaged citizens in some U.S. communities, and contributes heavily to programs that address environmental concerns, such as saving the world's tropical rainforests from destruction.

Needless to say, customers have been impressed with Ben & Jerry's concern for good causes, good communities, and of course with their good, creamy ice cream, considered by some to be the best in America. Their unique combination of idealism and ingenuity has made Ben & Jerry's Homemade what it is today. If you would like to find out more about Ben & Jerry's programs for peace and other projects, write to:

Ben & Jerry's
1% for Peace
P.O. Box 658
Ithaca, New York 14851

Source: *Ben & Jerry's Annual Report,* 1989.

QUESTIONS

First, scan the text until you find the answer. Then, underline the part of the text that contains the answer and then write the question number next to it. After you have located your answer in the reading passage, write short answers to the questions below in your own words.

1. What type of company is Ben & Jerry's Homemade?

____

____

2. Was Ben & Jerry's always a large corporation?

____

____

3. What is Ben & Jerry's philosophy regarding business?

____

____

4. What is the purpose of the organization 1% for Peace?

____

____

5. What other types of programs does Ben & Jerry's sponsor?

____

____

6. How have consumers responded to Ben & Jerry's Homemade?

____

____

RELATED READING 2

Read the article *The Body Shop* to prepare for the task that follows.

The Body Shop

In 1979, Anita Roddick, a 33-year-old English housewife and mother in Brighton, England, borrowed the equivalent of six thousand U.S dollars from a bank and used it to open a little shop stocked with shampoos, lotions, and creams. All of the products in the shop were made with natural, if exotic, ingredients from remote parts of South America. Roddick incorporated three of her greatest passions into her little business in the town of Brighton: her love of travel and foreign cultures, her fascination with traditional herbal formulas that promoted healthy skin and hair, and her great concern with saving the earth from environmental pollution and destruction.

Anita Roddick, along with her husband, Gordon, has been using the Body Shop not only as a place to sell natural cosmetics and cleansers but also as a clearinghouse for information on environmental problems and how to solve them. Next to the shampoos and facial creams, customers will find shelves stocked with reading material on the plight of inhabitants of the rain forest and of other endangered parts of the world. In addition, the shop has always taken a strong position against using animals to test cosmetic product safety. Roddick feels that if cosmetic products contain only natural ingredients and no harmful, artificial ones, such product safety testing, which causes untold suffering to animals, will no longer be necessary.

The packaging of Body Shop products is plain, simple, and recyclable. At the same time, it incorporates written information that promotes environmental awareness. In addition, the staff of the Body Shop must go through a special training program that helps them to become experts in the field of herbal cosmetics. The shop's management feels it's important to invest in a staff training program, as knowledgeable salespersons will take more of an interest in their work and can better serve customers.

The Body Shop has also felt an obligation to rechannel its profits back to those developing countries where its products originated by helping to set up members of communities there as suppliers. Through the Body Shop's program called Trade Not Aid, the economies of many of these communities has received a boost.

Today the Body Shop has grown from a small shop in an English town to a huge international corporation with over 141,000,000. U.S. dollars in sales. The Body Shop now operates numerous shops in thirty-seven different countries. The United States and Japan are the Body Shop's largest markets.

Source: Burlingham,B. "This Woman Has Changed Business Forever," *Inc.* June 1990.

WRITE YOU OWN QUESTIONS

Write your own set of questions to accompany the article, *The Body Shop*. Begin your questions with the words below. Then ask your partner to answer the questions without looking at the text:

1. Who ____?
2. When _______________________________________?
3. Where ______________________________________?
4. How ____?
5. Why ____?

WRITE A NEWS STORY

Imagine that you are writing an article for the local newspaper. What information about the Corporate School, Ben & Jerry's, and The Body Shop would be of particular interest to the people in your community? Write a piece of at least 200 words. Be sure to include at least five of the new vocabulary words and/or expressions you learned in the course of studying this segment. While writing your article, consider the following question:

> Do businesses have an obligation to help improve the society of which they are a part? Or is expanding an making profits the only real "business" of businesses?

INTERVIEW

Tell a business person in your community about the Corporate School, Ben & Jerry's, and the Body Shop. Ask this business person how he or she feels about the idea of running a business with "a social conscience." Write one or two paragraphs summarizing your interview with the business person and explaining his or her views on the subject. Use at least three or four of the new words and/or expressions that you learned in the course of studying the segment.

SUGGESTED READINGS

Collins, Marva & Tamarkin, Civia. *Marva Collins' Way*. Tarcher Publishers, 1982.

Cetron, Marvin & Gayle, Margaret. *Educational Renaissance*. St. Martin's Press,1982.

Harman, Willis & Hormann, John. *Creative Work: The Constructive Role of Business in a Transforming Society*. Knowledge Systems, Inc., 1990.

Holtz, Harvey, et. al. *Education and the American Dream*. Bergin & Garvey Publishers, 1989.

Kozol, Jonathan. *Savage Inequalities*. Crown Publishers, 1991.

Segment 3

Wild About Learning

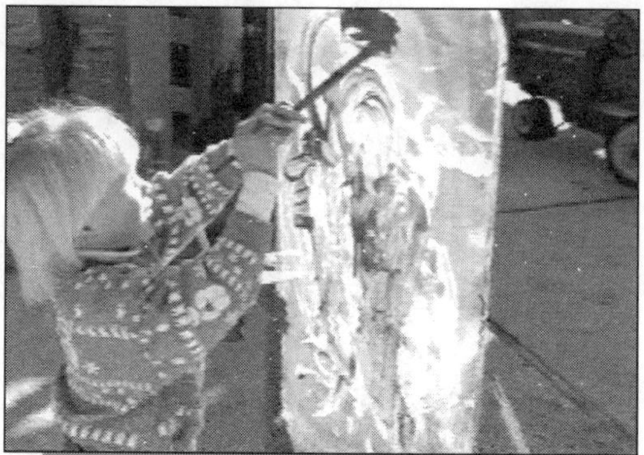

from *20/20,* 4/19/91
Runtime: 14:50
Begin: 00:21:24

Early childhood educator Bev Bos believes young children learn best when schools offer them fewer rules and more fun.

Previewing

KEY QUESTIONS

1. Should school be a "fun and play experience" for young children or a serious experience that prepares them for later learning?
2. What is the best age for children to begin learning basic academic subjects like reading, writing, and arithmetic? Why?
3. What are some of the most important things we can teach pre-school and elementary school children?

DISCUSSION QUESTIONS

1. How much freedom to experiment and try new things should young children be given in a nursery school or pre-school?

____

____

2. Do you think it is important to set down rules and restrictions in a school environment? If so, what types of rules should be set down?

____

____

3. Should the students or the teacher initiate learning activities? Why?

____

____

PREDICTION

Based on the title of the segment, *Wild About Learning,* the discussion you've had with your classmates, and your own knowledge, what information do you think will be included in this video segment?

1. ____

2. ____

3. ____

ESSENTIAL WORDS TO KNOW

The *italicized* words in the sentences below are used in the video. Read the sentences, then match the words with their meanings.

radical:	Her approach to education was considered *radical* because it was so non-traditional.
appalled:	She was *appalled* when she learned that the teacher did not care about his students.
encourage:	His ideas make learning more fun for children, but what he *encourages* most is for teachers to be less serious.
eccentric:	Her *eccentric* teaching method allowed children to yell and scream and run around in the classroom.
self-esteem:	The children learned *self-esteem* through their achievements and success.

______	1. radical	a. differing from what is considered normal
______	2. appalled	b. a healthy feeling about oneself
______	3. encourages	c. unorthodox; new and different
______	4. eccentric	d. horrified
______	5. self-esteem	e. help to promote

Global Viewing

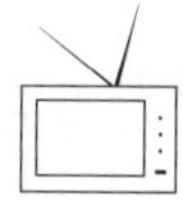

00:21:24-
00:36:14

CONFIRMING YOUR PREDICTIONS

Watch the entire video segment. As you watch, put a check (✔) next to any of your predictions that were mentioned.

Prediction 1 ______

Prediction 2 ______

Prediction 3 ______

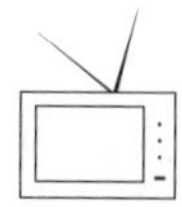

00:21:24-
00:36:14

MAJOR POINTS/MAIN IDEA

With a partner, watch the entire segment. Discuss some of the major points made in the program and write them below.

____

____

____

____

____

____

In your opinion, is one particular point the most important, or does one seem to include the other points? Write that point below in one sentence.

____

____

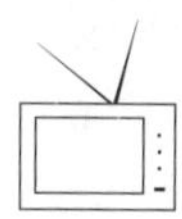

00:21:24-
00:36:14

THINK, WRITE, AND DISCUSS

Watch the entire segment and answer the questions below. If you wish, you may take notes on a separate sheet of paper. After you've finished, discuss your answers with the class.

1. What is Bev Bos's philosophy of early childhood education?

   ____

   ____

2. What is her major criticism of most pre-school programs?

   ____

   ____

3. How is Bev Bos's pre-school program different from most others?

____

____

4. What does Bev Bos believe is the most important thing for young children to learn?

____

____

5. Why does Bev Bos feel that freedom in a learning environment is so important?

____

____

Intensive Viewing

LISTENING CLOZE

Watch this part of the segment and fill in the missing words. Then compare your answers with those of other students.

00:22:13-
00:22:42

Barbara Walters: President Bush ________________ his new education program yesterday. It comes after a ____________ this week that some American companies are going ________________ because foreign workers are better-educated.

It's no ________ that American children have ____________ behind. Some say one ________________ is to start teaching them at an ____________ age, but Stone Phillips has found an energetic, somewhat ________________ woman whose message is, "Let the kids be kids." What ________ she mean by that? Watch.

NOTETAKING

00:24:07-00:35:56

As you watch the segment a second time, take notes to fill in the information below.

1. The three problems Bev Bos sees in every classroom:
   a. ____________________
   b. ____________________
   c. ____________________
2. The three times Bev Bos interferes in children's activities:
   a. ____________________
   b. ____________________
   c. ____________________
3. What experts say about Bev Bos's pre-school in Sacramento:
   ____
   ____
4. According to Sue Bredekamp, president of the Organization of Early Childhood Educators, when children are playing, they are: __________
5. Bev Bos says that you can't get kids ready for school, you've got to:
   ____
6. Besides children who does Bev Bos teach? __________
7. The three basics, according to Bev Bos:
   a. ____________________
   b. ____________________
   c. ____________________
8. What Bev Bos says about experience: __________
   ____
9. Why Bev Bos's ideas are not radical: __________
   ____

10. Age-appropriate learning is: ______________________________

____

11. A push-down curriculum is: ______________________________

____

12. At the Creme de la Creme Pre-School in Houston, children study:

____

____

13. Learning to read

a. What Roberta Babb says: ______________________________

____

b. What Bev Bos says: ______________________________

____

____

14. Concerning interruptions, Bev Bos says:

____

____

15. Two of the biggest reasons children fail:

a. ______________________________

b. ______________________________

16. What Hugh Downs and Stone Phillips say about:

a. Japan: ______________________________

____

b. Denmark: ______________________________

____

c. the United States: ______________________________

____

WHAT'S DIFFERENT?

00:24:07-
00:35:56

Look at the notes you took in the exercise above. How is the Creme de la Creme Pre-school different from Bev Bos' pre-school? Which do you think is a better program for children? Use the space below to write your opinions. Then compare your opinions with that of another student. If you need to, you may watch the segment again.

____

____

____

____

Language Focus

MULTIPLE WORD VERBS

A multiple word verb is composed of a verb and one or more prepositions. Together these words have a specific meaning. The following sentences are from the segment. Match the *italicized* multiple word verbs with their synonyms from the list below.

a. to be less serious
b. change
c. expose to
d. remove
e. reject

______ 1. The director of this school, Mr. Thaxton, doesn't like what he sees. Today he's brought in a visitor he hopes will *turn* things *around*.

______ 2. By calling it "pre-school" we *set* ourselves *up* for criticism.

______ 3. If kids aren't social by age six, other kids *push* them *away*.

______ 4. Kids don't catch colds from *taking* their clothes *off*.

______ 5. What she encourages most of all is for teachers *to lighten up*.

TWO-WORD VERBS

In the sentences on page 31, we find separable two-word verbs. If a two-word verb is separable, we can separate the verb and the prepositions with an object noun. For example, we can say:

Kids don't catch cold from *taking* their clothes *off*.

Or we can say:

Kids don't catch colds from *taking off* their clothes.

If we replace the object noun with a pronoun, the pronoun must be placed between the verb and the preposition; it cannot be placed after the preposition. For example,we can say:

Kids don't catch colds from taking *them* off.

But we cannot say:

Kids don't catch colds from taking off *them*.

The sentences below contain separable two-word verbs. Replace the *italicized* object nouns with pronouns and place them correctly in each sentence. The first one has been done for you.

1. He picked up *English* when he worked with Americans.

   He picked **it** up when he worked with Americans.

2. Bev Bos has been turning around *teaching methods* for the last decade.

   ____

   ____

3. Bev Bos set up *her school* thirty years before the television program about her teaching methods was aired.

   ____

   ____

4. The parents picked up *their children* after school.

   ____

   ____

5. The child picked up *the toy*.

   ____

   ____

VISUAL CONTEXT CLUES

Watch this part of the segment. Based on the images you see, write your own definition of the *italicized* words below.

1. "Blowing *bubbles* every day—is that a learning experience?"
   "Yes! How come a *bubble* is always round? Because of the air pressure."
   bubble: ______________________________ 00:26:35-00:26:40

2. She's *on the road* every week. 00:27:32-00:27:40
   on the road: ______________________________

3. "Is there anybody here who doesn't have a *toothpick* and a *wire*? 00:27:56-00:28:20
   Wrap [the wire] around the toothpick down low, and it will balance no matter where you put it."
   wire: ______________________________
   toothpick: ______________________________
   wrap: ______________________________

4. "Wonder! Wanting to know! Look at you people with this toothpick and this wire—I mean you've just *gone nuts*!" 00:28:15-00:28:38
   gone nuts: ______________________________

Postviewing

DO YOU AGREE OR DISAGREE?

Work with a small group of students and discuss whether you agree or disagree with the statements below. Be prepared to justify your opinion. Choose one student to be the leader of the group. The group leader can record and later report the group's opinions to the rest of the class.

1. Children should begin learning academic subjects, such as reading, writing, and arithmetic, at as young an age as possible—perhaps at age three or four.
2. Bev Bos' teaching philosophy gives children too much freedom.
3. The teacher should always decide what is best for young children to learn—children should not initiate their own learning activities.
4. Schools like the one Bev Bos runs would work well in my country.
5. If children are given a lot of freedom in their education, they won't have enough discipline later in their school careers.
6. Bev Bos is right when she says, "Experience is not the best teacher—it's the only teacher."

RELATED READING

Read the article *Content-Based Learning* to answer the questions that follow.

Content-Based Learning

Most people can remember a time in their lives when they learned something almost by accident, that is, without consciously trying to learn it. Often this kind of learning happens incidentally, while we are focusing our attention on learning something else. For example, many people learn a number of English words not by memorizing them or studying them directly, but just by doing something they enjoy, like listening to popular songs that contain them. Similarly, some people learn words in Chinese or Japanese not by studying those languages directly, but by studying martial arts, such as kung fu or aikido, in which Chinese or Japanese terms are used. Foreign students in the United States often learn the system of measurement simply by having to shop and cook for themselves, activities that require them to learn units of measure that include cups, quarts, and ounces.

Many educators believe that such incidental learning, generally called content-based learning, is the best way to grasp the underlying rules of a system. Proponents of content-based learning argue that people can best understand and internalize the logic of a system's basic rules by coming upon them in the context of that system—not by studying and memorizing them in their abstract form or out of context. Therefore, proponents of content-based learning feel that grammatical rules can best be grasped indirectly through the process of reading and writing, rather than by rote memorization. Likewise, they feel that mathematical rules can best be learned by working out real-life problems where such rules are needed to find solutions to the problems. In many cases, our own learning experiences may support the idea that we have a deeper understanding of rules inferred from context than of those memorized out of context.

QUESTIONS

Write short answers to the questions below. Then underline the part of the article that justifies your answer to the question and write the number of the question next to it.

1. What can most people probably remember?

   ____

   ____

2. What are some examples of incidental learning?

   ____

   ____

3. What do many educators believe?

____

____

4. Why do they believe this?

____

____

5. What would a supporter of content-based learning believe regarding learning the grammatical rules of a language?

____

____

COMMITTEE WORK

In this segment you have learned about many different philosophies of early childhood education. Form small groups and come up with an early childhood education program of your own. You may incorporate ideas from the philosophies you now know of, or you may create an entirely new program. Present your ideas to the class in an oral report.

PROJECTS

Choose one of the following exercises.

IDEA # 1

Interview one or more teachers about their philosophy of educating young children. When preparing your list of interview questions (which you may do alone or with a partner) be sure to use four or five of the vocabulary words studied in this chapter. After the interview, compare their ideas with those of Bev Bos. How are they the same? How are they different? Write a paper in which you report the similarities and differences in their views.

Idea #2

Writing teachers who support the notion of content-based learning feel that the best way for students to learn to write is to involve themselves in a long-term writing project, often one with a personal theme (as this is thought to be of greatest interest to the student). In this type of writing, called process writing, students are asked to focus on expressing their ideas rather than on achieving grammatical perfection. It is believed that through the process of writing, students will make discoveries about the rules that underlie the language they are using, and, in time, that their writing will naturally improve.

Test out the notion of process writing. Start keeping a journal in English and write in it several times a week. Ask your teacher to look at it from time to time. Both you and your teacher can assess whether process writing is helping you to gain a greater command of the English language.

SUGGESTED READINGS

Elkind, David. *Miseducation*. Knopf, 1989.

Lefkowitz, N. *From Process to Product*. Regents/Prentice-Hall, 1990.

Montessori, Maria. *The Secret of Childhood*. Ballantine Books, 1966.

Neill, A.S. *Summerhill: A Radical Approach to Child Rearing*. 1984.

Standing, E.M. *Maria Montessori: Her Life and Work*. Plume, 1957.

Segment 4

The Food of the Future

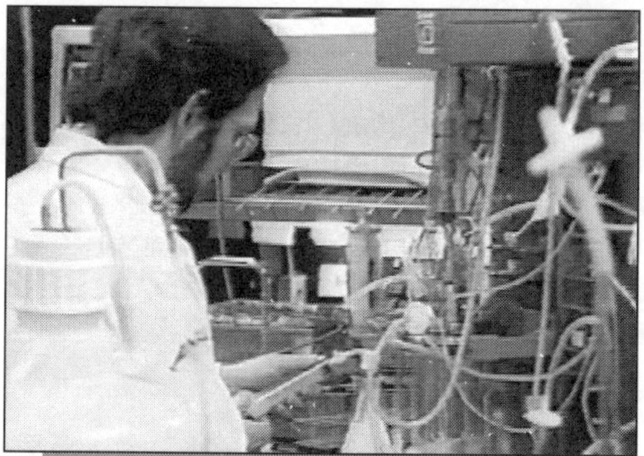

from *World News Tonight*, 3/21/88
Runtime: 4:02
Begin: 00:36:21

Rapid scientific developments in this century promise to change our notions of eating.

Previewing

KEY QUESTIONS

1. Is the food that people eat today different from the food they ate in the past?
2. Has the development of new food products improved people's nutrition and eating habits? If so, how? If not, how has it hurt?
3. In the future, there will be less farmland. How might that fact affect the way food is produced?

DISCUSSION QUESTIONS

1. Can you think of any examples of how science has already changed the food we eat? Have these changes been beneficial to us?

____

____

2. How do you think the food that future generations eat will change? Will it change for the better or for the worse?

____

____

3. In recent years, scientists have been learning more and more about how to change the genetic structure of plants and animals. Do scientists have the right to tamper with nature in this way?

____

____

PREDICTION

Based on the title of the segment, *The Food of the Future*, the discussion you've had with your classmates, and your own knowledge, what information do you think will be included in this video segment?

1. ____

2. ____

3. ____

4. ____

ESSENTIAL WORDS TO KNOW

The *italicized* words will be used in the video. Read the words as they are used in the following sentences and try to come up with your own definition.

1. The United States has traditionally had an *abundance* of food; it produces enough food both for its own use and for export to other countries.

   abundance: ____

2. One problem farmers face is *erosion*, which can occur when windstorms blow their farmland away.

   erosion: ____

3. When a plant's *genetic* structure is changed, its characteristic features may be different.

   genetic: ____

4. Most plants and animals are made up of millions of *cells*, the essential element of every living thing.

   cell: ____

5. In some areas of the country, *livestock*, rather than crops, is the main food product.

   livestock: ____

6. *Agriculture* is the main industry in areas that have rich soil and enough water and land to grow crops and raise animals.

   agriculture: ____

Global Viewing

CONFIRMING YOUR PREDICTIONS

00:36:21-00:40:23

Watch the entire video segment. As you watch, put a check (✓) next to any of your predictions that were mentioned.

Prediction 1 ______

Prediction 2 ______

Prediction 3 ______

Prediction 4 ______

MAJOR POINTS/MAIN IDEA

00:36:21-00:40:23

After you watch the segment, work with a partner and write down a few of the major points made in the program.

____

____

____

____

____

____

Now compare your major points with those of other students in the class. Ask one of the students to write on the chalkboard all of the major points students have identified. Then, with the other members of the class, decide which of the points is the most important and write it below.

____

____

THINK, WRITE, AND DISCUSS

00:36:21-00:40:23

Watch the entire segment and answer the questions on page 40. If you wish, you may take notes on a separate sheet of paper. After you've finished, discuss your answers with the class.

1. How will the food of the future be different from the food of today?

____

____

2. What importance will vegetables like the winged bean and soybean have in the future?

____

____

3. In the future, why will vegetables like the winged bean and the soybean be important?

____

____

4. How do scientists intend to achieve many of the future improvements in vegetables, fruits, and fish, and meat?

____

____

5. Are scientists improving upon nature?

____

____

Intensive Viewing

LISTENING CLOZE

Watch this part of the segment and fill in the missing words. Then compare your answers with those of another student.

00:36:25-00:36:43

Peter Jennings: We have a ____________ tonight about the food we are __________ to eat. We've all __________ the saying "you are ________ you eat," and recently, we've heard about the __________ toward more __________________ in food, a fat ____________________ for example, or a new ____________________ sugar. And so we really _______________ what kind of food we would be eating in the next _____________.

NOTETAKING

00:36:44-
00:37:51

The notetaking form below has been partially filled in for you. First read through the notes below so that you know what to listen for. Then, as you view the video, fill in the missing information. When you finish, compare your notes with others to check your comprehension. Then watch the video again, filling in any information you missed the first time.

I. Foods undergoing most radical changes: ______________________

   A. Differences

      1. will develop more sugar

      2. ______________________________

      3. ______________________________

II. Tomatoes of the future

   A. Genetic structure manipulated to ______________________

   ____

   B. Tomato products made without ______________________

III. Tomato cells

   A. Similarity to whole tomatoes

      1. taste

      2. ____________________

   B. Products that may use tomato cells

      1. ____________________

      2. ____________________

   C. Advantage: ______________________________

GETTING THE DETAILS

00:37:51-
00:39:58

Watch this part of the segment and circle the letter of the item that **does not** correctly complete the following sentences.

1. Food factories will grow vegetables ________.
   a. using cells
   b. without soil
   c. by bathing their roots only in water and minerals
   d. on assembly lines

2. The beef of the future will come from steers that ________.
   a. are cloned
   b. are bred to grow faster
   c. are fatter
   d. need less feed
3. Meat substitutes being developed include ________.
   a. new types of fish that grow faster and taste milder
   b. soybean products that look and taste like meat
   c. products that don't have a high cholesterol content
   d. tomato cells that don't come from tomato plants
4. The healthier food trend may encourage the development of ________.
   a. cereal with aspirin
   b. meat with no cholesterol
   c. ice cream with heartburn medicine
   d. medicinal foods

Language Focus

COMPARATIVE ADJECTIVES

Several of the sentences in the program segment contain comparative adjectives; in the sentences below, the tomatoes of the present are compared to the tomatoes of the future. For example:

> [Tomatoes] will develop more sugar and be *tastier* . . .
>
> Their genetic structure has been manipulated to make them *less likely* to spoil.

Now look at the chart below that summarizes how to form comparative adjectives.

	ADJECTIVES WITH ONE OR TWO SYLLABLES	ADJECTIVES WITH THREE OR MORE SYLLABLES
more	adjective + er (if adj. ends in *y*, change to *i* before *-er*)	*more* + adjective
less	*less* + adjective	*less* + adjective

Using the adjectives below and the information you learned in the segment, make sentences that compare the food of the future with the food of the present. Use each word only once. The first one has been done for you.

dependent easy efficient healthy lean sweet

1. The tomatoes of the future will be sweeter than today's tomatoes.
2. The beef of the future will be ____________ than the beef of today.
3. Farming methods of the future will be ________ ___________________ than the farming methods of today.
4. Farming methods of the future will be ________ ___________________ on available farmland.
5. For some people, the ice cream of the future will be ____________ to digest than the ice cream of today.
6. In general, the food of the future will be ___________________ than the food of today.

VOCABULARY IN CONTEXT

Based on the context clues in the following sentences, guess the meanings of the *italicized* words. First circle the words that helped you to guess the meaning. Then write a definition in the space provided.

1. What's more, this technology can help *alleviate* farmland shortages, as scientists can grow the equivalent of a whole acre of tomatoes in just a few flasks.

   alleviate: ____

2. This squash is grown without soil, *hydroponically.*

   hydroponically: ____

3. These steers are *cloned,* identical copies produced from the genes of one genetically superior animal.

   cloned: ____

4. The healthier food trend in the next century will include *medicinal* foods. Drug companies are incorporating aspirin into cereal and heartburn medicine into ice cream.

   medicinal: ____

Postviewing

DO YOU AGREE OR DISAGREE?

Work with a small group of students. Discuss whether you agree or disagree with the following statements and why. Choose one student to be the group leader. The leader can report the opinions of each member of the group to the rest of the class.

1. It's a good thing that people will eat less meat in the future.
2. Scientists should not tamper with nature and grow food in such unnatural ways.
3. Scientists are developing better methods of growing food than nature did.
4. It's very sad that in the future, the traditional farm will no longer exist.
5. The cloning process will play a very big role in many aspects of life in the future, and this will beneficial to everyone.

RELATED READING 1

In the program segment, we are told that "there are new soybean meat substitutes . . . that can be shaped to look like anything from hot dogs to veal cutlets." Read the short article below about a soybean product called tofu, or bean curd, that has been used for centuries in many parts of the world and promises to be used even more in the future. Then answer the questions that follow.

Tofu: The Ancient Food of the Future

In the last several decades, the western world has been rediscovering the soybean, a plant whose use pre-dates recorded history in Asia. For thousands of years, the Chinese people have extolled virtues of the soybean, considering it one of the five sacred grains (along with rice, millet, barley, and wheat). The soybean is an almost perfect protein that contains seven of the eight amino acids. Since it is also cheap to produce, free of cholesterol, and versatile in its uses, the soybean could provide food to the world's growing population. Besides nourishing human beings, soybeans can be fed to a variety of other animals ranging from livestock to silkworms.

In the 1930s the United States and Canada undertook a research project in which they learned to breed an especially nutritious soybean with a high oil content. Since then the United States has become the world's largest producer of soybeans. Although the U.S. exports much of its soybean produce, in recent years soybean curd, or tofu, has begun making inroads into the American diet. Health-conscious Americans are adding tofu to salads, and casseroles and using it as a meat substitute. In the health food stores of major cities, it's now possible to buy tofu burgers and even tofu hot dogs.

Source: *Barnes & Nobles Encyclopedia,* vol 18, Grolier, 1991.

QUESTIONS

Write short answers to the questions below. Be sure to underline those parts of the above passage that justify your answers.

1. What has been happening in the western world?

2. Where were soybeans originally used?

3. Why do you suppose the soybean was considered a "sacred grain"?

4. When did the western world begin to take a special interest in soybeans?

5. What happens to most of the soybeans grown in the U.S.?

6. What have health-conscious Americans been doing in recent years?

7. What is it now possible to buy in the United States?

8. What importance do you think soybeans will have as a food product in the future?

RELATED READING 2

Read the article below to prepare for the task that follows.

Epcot Center

If you are in the United States or if you plan to visit it one day, you can visit a farm of the future. The farm of the future is one of the many exhibits at Epcot Center, an educational amusement park that shares a large tract of land with the world famous Disney World in Orlando, Florida. Epcot Center's exhibit shows that farms of the future will bear little resemblance to the farms of today; rather, the farms of the future will be more like factories, producing food assembly-line style. These "farm factories" will be able to produce food far more quickly than could ever be imagined with the farming methods of today. The farms of the future will better meet the food needs of the world's growing population.

QUESTIONS

Write questions about the above reading passage. Ask your partner to answer the questions without looking at the text. Begin your questions with the words below.

1. What____________________________________?

2. Where___________________________________?

3. How_____________________________________?

4. Why_____________________________________?

NOTABLE QUOTES

At the beginning of the program segment, the moderator, Peter Jennings says, "We've all heard the saying, *'You are what you eat'*." Write a paper explaining what the statement means and agreeing or disagreeing with it.

PROJECT

Write your reaction to one of the statements below. Develop your ideas in one or more paragraphs.

1. The new high-technology methods of farming will greatly change humankind's relationship to nature.
2. Because of the increasing emphasis on speed and efficiency in modern life, time-consuming practices such as cooking and even letter-writing will become "lost arts" in the future.
3. One day far in the future, people will view the practice of eating meat as archaic and perhaps even barbaric.

SUGGESTED READINGS

Campbell, Keith. *Food for a Future: How Agriculture Can Meet the Challenge.* University of Nebraska, 1979.

Carmen, Ira H. *Cloning and the Constitution: An Inquiry into Government Policy-Making and Genetic Engineering*. University of Wisconsin, 1986.

Moore, Frances Lappe. *Diet for a Small Planet.* Ballantine Books, 1971.

Robbins, John. *Diet for a New America.* Stillpoint Press, 1987.

Wynne-Tyson. *Food for a Future: How World Hunger Could Be Ended by the 21st Century.* Borgo Press, 1989.

Segment 5

Virtual Reality

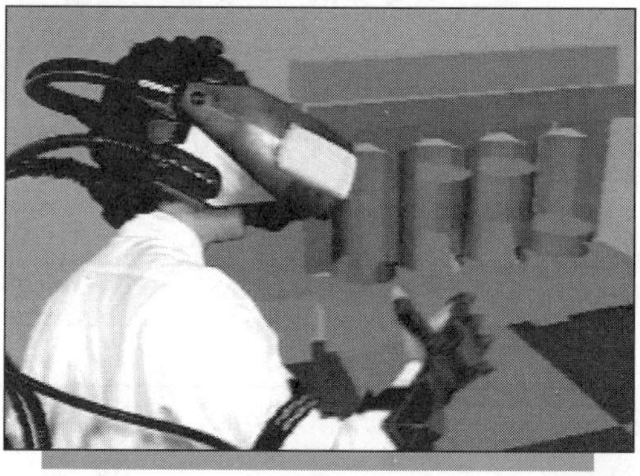

from *American Agenda*, 9/19/91
Runtime: 11:13
Begin: 00:40:31

Imagine going inside a TV screen and magically becoming part of a program—or a video game.

Previewing

KEY QUESTIONS

1. What developments in computer technology have changed the way people live and work?
2. How have home entertainments such as television, video recorders, and video games affected people's lives?
3. It has been said that "Technology is a double-edged sword." What does that statement mean?

DISCUSSION QUESTIONS

1. How have new developments in computers and technology directly affected your life? Give examples.

   ____

   ____

2. Are advances in technology always beneficial? Think of some cases where inventions resulting from advances in technology have also had some negative effects on human society.

   ____

   ____

3. How will further advances in computer technology continue to change the world?

____

____

PREDICTION

Based on the title of the segment, *Virtual Reality*, the discussion you've had with your classmates, and your own knowledge, what information do you think will be included in this video segment?

1. ____
2. ____
3. ____
4. ____

ESSENTIAL WORDS TO KNOW

The following words will be used in the video. Read the words as they are used in the following sentences and try to come up with your own definition.

1. Using computers to create graphics and sounds, *virtual reality* makes the viewer believe he or she is in another world.

   virtual reality: ____

2. Virtual reality may seem unbelievable, but it's not as *far-fetched* as it sounds.

   far-fetched: ____

3. *Three-dimensional* images are created using technology that fools the viewer's mind into perceiving depth.

   three-dimensional: ____

4. *Submerged* in computer graphics and sounds, the viewer is cut off from reality and completely involved in his or her own world.

   submerged: ____

Global Viewing

CONFIRMING YOUR PREDICTIONS

00:40:31-00:51:44

Watch the entire video segment. As you watch, put a check (✓) next to any of your predictions that were mentioned.

Prediction 1 _____

Prediction 2 _____

Prediction 3 _____

Prediction 4 _____

MAJOR POINTS/MAIN IDEA

00:40:31-00:51:44

Work with a partner and write down some of the major points made in the program.

Does one point seem to be the most important, or include the others? Write the main idea of the program below.

THINK, WRITE, AND DISCUSS

00:40:31-00:51:44

Watch the entire segment and answer the questions below. If you wish, you may take notes on a separate sheet of paper. After you've finished, discuss your answers with the class.

1. What is virtual reality?

2. Who can use virtual reality?

____

____

3. How could virtual reality benefit society?

____

____

4. How might virtual reality harm society?

____

____

Intensive Viewing

LISTENING CLOZE

00:43:19-
00:43:41

The passage below is from the beginning of the video segment. Watch this part of the video segment again and fill in the blanks with the missing words.

Jay Schaedler: As with most VR systems, Waldern's uses stereoscopic images ______________ onto tiny ______________ inside the viewer's ____________. A powerful computer then ______________ the graphics and ____________ needed to convince the viewer he's in another world. The ______________ is completed when motion sensors in Waldern's glove and ____________ allow him to begin ______________ with whatever he sees.

WHO USES VIRTUAL REALITY?

00:43:55-
00:47:10

Look over the chart on page 53. Then watch the next part of the segment and fill in the chart with the appropriate information. Watch the video as many times as necessary. When you finish, compare your notes with others to check your comprehension.

User	Use	Implementation	Benefit
NASA	recreating different worlds		
Urban planners		not given	
Architects			early problem solving
Medicine		turning a CAT scan into 3-D model of the patient's body	

THE FUTURE OF VIRTUAL REALITY

00:49:21-00:51:15

As you watch the program a second time, take notes to fill in the information below. After viewing, check your answers with those of another student.

1. Virtual reality had its roots in ______________________________

   ____

2. Howard Rheingold says that with VR, the line between playing a game and actually destroying people is ______________________________

   ____

3. Rheingold says virtual reality will force us to ask, ______________________________

   ____

4. Virtual reality might be a turning point for the human race because____

   ____

5. According to the program, VR systems cost______________________________

   ____

Language Focus

CLOZE EXERCISE

Below is a passage of the program's narration. Write the correct forms of the
00:40:15- words below in the blank spaces. Then watch the segment to check your
00:41:42 answers.

be have see do leave write

Sam Donaldson: It was more than a hundred years ago that Lewis Carroll __________ about Alice's trip through the looking glass. Now that fiction ______ become a reality — or you might say, a virtual reality — because that's the name of a new computer technology that many believe will revolutionize the way we live. Virtual reality lets you travel to places you've never ________ , see things you've never ________ , do things you've never ________ — without ______________ the room.

VISUAL CONTEXT CLUES

Watch this part of the segment and use the visual clues to describe the *italicized* words.

00:47:38- 1. I've got a *joystick,* I've got a *helmet.*
00:47:42 *helmet:* ____

joystick: ____

00:47:45- 2. To fire the gun, you simply press the *trigger.*
00:47:48 *trigger:* ____

00:48:24- 3. I'm a little worried by that flying *pterodactyl* at this point.
00:48:35 *pterodactyl:* ____

00:49:06- 4. The visual display is *crude*. There are these *crude,* cartoon-like characters.
00:49:09 *crude:* ____

VOCABULARY IN CONTEXT

Based on the context clues in the sentences on page 55, guess the meanings of the *italicized* words. First circle the words that helped you guess the meaning. Then write a definition of the word in the space provided.

1. Virtual reality *straddles* the foggy boundary between fantasy and fact.

   *straddles:*____

2. Imagine a place and you'll be able to step into it. *Conjure* up a dream and you'll be able to fly through it.

   conjure: ____

3. He's *launched* one of the first companies to mass-produce VR systems.

   launched: ____

4. VR techniques have been used to make a three-dimensional model of the planet Mars. There are, of course, more *down-to-earth* applications. VR models of urban landscapes are allowing urban planners to redesign Main Street without ever leaving home.

   down to earth: ____________________________________

5. We're now reaching a point where the simulations are so realistic that the line between playing a game or a simulation and actually blowing people up is becoming *blurred.*

   blurred: ____

STRUCTURE FOCUS: *BE ABLE TO*

The meaning of the expression *be able to* is synonymous with *can*. Students sometimes experience confusion conjugating *be able to* in the various tenses. Only the verb *be* in this expression carries tense. In the exercise below, note the tense of each sentence and fill in the appropriate tense of the verb *be*.

1. Presently, in Japan, consumers _______ able to try out their home decorating plans using virtual reality.
2. Jay Schaedler, the news broadcaster, ____ able to experience the world of virtual reality by playing Nightmare, the video game that John Waldern created.
3. In recent years, the military __________________ able to use virtual reality technology in flight simulation.
4. In the future, people _____________ able to have easy access to virtual reality systems.
5. If virtual reality technology were more affordable at the present time, many more people __________________ able to try it.

WORD FORMS

The following words were used in the segment. Complete the chart below by adding the other forms of the given words.

	NOUN	VERB	ADJECTIVE
1.	reality	(none)__________	__________
2.	exploration	__________	__________
3.	__________	(none)__________	dimensional
4.	__________	__________	imaginary
5.	application	__________	__________
6.	pioneer	__________	__________
7.	__________	__________	dangerous
8.	__________	__________	sensitive

Postviewing

DO YOU AGREE OR DISAGREE?

After you have completed your discussion, choose a group leader to share the group's opinion with the rest of the class.

1. Virtual reality will be a turning point for the human race.
2. Virtual reality will do more harm than good to humanity.
3. The best uses of virtual reality will be in the fields of medical and military technology.
4. Some day, virtual reality will make other forms of entertainment, such as television and movies, obsolete.

RELATED READING

Read the article *Jules Verne: From Science Fiction to Science Fact* and answer the questions that follow.

Jules Verne: From Science Fiction to Science Fact

Humankind often dreams of what is beyond its ability to do. Dreamers can, however, often pave the way for doers — although the time lapse between the dreaming and the doing may be decades or even centuries. Artists and writers, the world's great dreamers, are often prophetic in their imaginary visions. Certainly this was true in the case of the first great French writer of science fiction, Jules Verne.

Verne lived from 1828 to 1905, when inventions such as the submarine and the spaceship seemed inconceivable. Yet Verne wrote what became a very famous novel, *From the Earth to the Moon*, about an expedition to outer space. He also wrote *Twenty Thousand Leagues Under the Sea* about submarine travel. His other works touched on the subjects of parapsychology and time travel.

Humanity's imagination and intuition work far faster than its capacity to produce a plan that will turn the product of human fantasies into realities. Although the connection between imaginative and rational thought is mysterious, it is clearly significant enough that the visions of the future described by Verne and other "prophetic" artists should be given serious consideration.

Source: *Encyclopedia Britannica*, Vol. 12. Encyclopedia Britannica, Inc., 1992.

QUESTIONS

Scan the text for information that answers the questions below. Underline those portions of the text that justify your answers and write the corresponding question numbers next to them. Then ask your partner to answer the questions without looking back at the text.

1. What is the relationship between "dreaming" and "doing"?

____

____

2. Who are the world's great dreamers?

____

____

3. Who is an example of one of these dreamers?

____

____

4. What did Jules Verne write?

____

____

5. Tell something about the connection between rational and imaginative thought.

____

____

CREATE YOUR OWN VIRTUAL REALITY

Come up with your own use for virtual reality. Work with a group and brainstorm. Think of some virtual reality applications that weren't mentioned in the video segment. For example, you might want to consider how virtual reality can be used in travel, history, psychotherapy, etc. Decide which virtual reality application you feel is best and design a plan. Present your plan to the class.

FROM SCIENCE FICTION TO FACT

The program on *Virtual Reality* mentions that a science fiction movie called *Tron* showed a virtual reality video world nearly a decade before such a thing existed. In many cases, science fiction movies and stories tell us of marvelous inventions of the imagination that later become realities. Work on your own or with the members of your group to think of as many such stories or movies as you can. Then develop your brainstorming into an oral report that you can deliver to the class in which you use at least three of the vocabulary words studied in this chapter.

NOTABLE QUOTES

Prepare an oral or written report explaining your reaction to the quotation below:

"What kind of reality are our grandchildren going to accept? Is the natural world with real trees and real rivers going to be there, or are we going to live in 800-story high-rises all plugged into our virtual realities in which we walk through artificial forests? This might be a turning point for the human race. We might be going inward toward an artificial experience."

— Howard Rheingold

RADIO PROGRAM

Option #1

Work with a small group of students and make a "radio program" composed of student reactions to the video segment *Virtual Reality*. Each member of the group can prepare a statement in which he or she comments on a different aspect of the segment. Decide in which order to put the comments. Take time to rehearse your script, correcting sentences as necessary. You may want to ask your teacher to check your written script before putting it on tape. In your tapescript, be sure to use some of the words and phrases that you learned from watching the program. After you complete your program, play it for the rest of the class. If you wish, you may have a class contest to choose the best program.

Option #2

You may wish to make a more spontaneous talk-show style program. In this case, it will not be necessary to write your script first, but simply to decide on questions the interviewer will ask during a round-table discussion you will record. Be sure to incorporate into the questions, and also into the responses, several of the vocabulary items studied in this chapter.

Afterward, the group can monitor the tape for correct use of vocabulary, pronunciation intonation, and grammar. You may wish to make re-make your taped program before playing it for the rest of the class. You may also wish to hold a class contest for the best radio program.

SUGGESTED READINGS

Hall, Stephen. *Mapping the Next Millennium*. Random House, 1991.

Krueger, Myron. *Artificial Reality*. Addison-Wesley, 1991.

Laurel, Brenda. *Computers as Theater*. Addison-Wesley, 1991.

Rheingold, Howard. *Virtual Reality*. Summit Books, 1992.

Segment 6

Conquering Fear at Camp Courage

from *20/20*, 9/15/89
Runtime: 17:25
Begin: 00:51:52

Dr. Jack Moncrief started Camp Courage, a place where seriously ill children can get away from their worries.

Previewing

KEY QUESTIONS

1 What types of medical technology help people to cope with their physical limitations?
2. How does an individual's physical limitations affect his or mental attitude?
3. What can a person do to overcome the fear associated with a physical limitation?

DISCUSSION QUESTIONS

1. Do children who are seriously ill have to lead limited lives? Why or why not?

____

____

2. What can be done to make a child who has a serious illness lead a more fulfilling and enjoyable life?

____

____

3. What are doctors' responsibilities toward the mental/physical recoveries of their patients?

____

____

PREDICTION

Based on the title of the segment, *Conquering Fear At Camp Courage*, the discussion you've had with your classmates, and your own knowledge, what information do you think will be included in the video segment?

1. ____
2. ____
3. ____

ESSENTIAL WORDS TO KNOW

The *italicized* words in the sentences below are used in the video. Read the sentences, then match the words on page 63 with their meanings.

impurities:	The *impurities* in the water were caused by the rusty lead pipes.
summer camp:	At the *summer camp* where they stayed, the children participated in many activities, including swimming, baseball, and hiking.
transplant:	The child with kidney disease lived because doctors were able to *transplant* a kidney from his mother's body to his own.
conquer:	She proved that she had *conquered* her fear of heights by climbing to the top of a mountain.
chronically ill:	The *chronically ill* patient suffered recurrences of her illness every five months.

______ 1. impurities	a. to overcome or defeat
______ 2. summer camp	b. a surgical procedure that transfers an organ from one person to another
______ 3. transplant	c. foreign substances that can be harmful to ones health
______ 4. conquer	d. suffering symptoms of a particular illness at regular intervals
______ 5. chronically ill	e. a place for children to vacation

Global Viewing

CONFIRMING YOUR PREDICTIONS

00:51:52-01:08:37

Watch the entire video segment. As you watch, put a check (✓) next to any of your predictions that were mentioned.

Prediction 1 _____

Prediction 2 _____

Prediction 3 _____

Prediction 4 _____

MAJOR POINTS/MAIN IDEA

00:51:52-01:08:37

Work with a partner and write below some of the major points made in the program.

____

____

____

____

____

____

Compare your list of program points with those of other students in the class. Are your points the same or different? One student should write on the board all of various points chosen by students. Is one point the most important? Does any one point include the others? Could this point be considered the main idea of the program? If so, write it below.

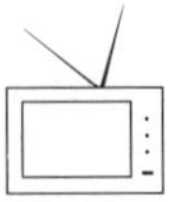

00:51:52-
01:08:37

THINK, WRITE, AND DISCUSS

Watch the entire segment and answer the questions below. If you wish, you may take notes on a separate sheet of paper. Then discuss your answers with the class.

1. What medical problem did Anthony Austin have, and how did his mother deal with it?

2. Why and how did Dr. Moncrief start the Moncrief Mountain Ranch (Camp Courage)?

3. What does Camp Courage offer to children?

4. What kinds of medical problems do the children at Camp Courage have?

5. Why do the camp counselors and Dr. Moncrief encourage the children to go down a 90-foot mountain cliff?

Intensive Viewing

LISTENING CLOZE

00:53:02-00:04:07

Watch this part of the segment and fill in the missing words. Then compare your answers with those of another student.

Hugh Downs: In the rolling hill country of southern Virginia, a young boy struggles to learn the piano. But eight years ago Anthony Austin was ____________________ for his life. He was born with ______________ that didn't work, and that used to mean certain __________ for an infant.

Peggy Austin: We'd never been on a __________ before, and we put you in a ______________ basket. We did.

Downs: But Anthony's mother, Peggy, came across a magazine article that __________ his life. It described a revolutionary new dialysis technique that had never been used before on a ______________ baby.

Mrs. Austin: And we took you to Austin, Texas, and you were very, very ________. Very sick. So they put you right into ______________, and you've had several ___________________, so you know what that's all about.

Downs: The new ________________ technique uses a simple plastic bag and tubing connected to the abdominal cavity to cleanse the system of impurities. It's ____________________ was Dr. Jack Moncrief in Austin, Texas.

TRUE OR FALSE

00:56:42-00:57:37

Watch the segment and decide whether the statements are true or false. Write *T* or *F*.

______ 1. Dr. Moncrief helped to invent a new dialysis technique.

______ 2. Dr. Moncrief strongly encouraged Anthony's mother to allow her son to have the operation.

______ 3. Dr. Moncrief's invention saved thousands of lives.

______ 4. The invention was a medical and financial success.

______ 5. Dr. Moncrief's co-inventor was the founder of the summer camp.

JACK MONCRIEF'S INVENTION

00:57:59-
01:04:27

Listen to the description of the CAPD dialysis technique developed by Dr. Moncrief and administered to camper Sathya Osterhaus. Pay special attention to the key words below and take notes. After viewing the segment, work with a partner and use your own words to write a description of the technique. Then answer the questions that follow.

KEY WORDS

dialysis fluid	tube
drain	impurities

____

____

____

____

1. What were the complications that resulted from the procedure in Sathya's case? ____

   ____

2. What did Dr. Moncrief recommend be done to deal with the complications? ____

   ____

THE CHILDREN

00:57:59-
01:04:27

Watch this part of the segment and answer the questions below.

1. According to Jack Moncrief, what happens when children face death so early?

   ____

   ____

2. What activities do the children learn at the summer camp?

   ____

   ____

3. What opportunity does the camp offer to children?

   ____

   ____

4. What does Jack Moncrief want the children to do at the camp?

____

____

5. How has Bryan changed because of the camp?

____

____

Language Focus

WORD FORMS

The following words were used in the segment. Complete the grid by adding the other forms of the given words.

	NOUN	VERB	ADJECTIVE
1.	____________	____________	revolutionary
2.	inventor	____________	____________
3.	medicine	____________	____________
4.	____________	____________	protective
5.	____________	to conquer	____________

PRACTICING THE PAST PERFECT

In the segment Dr. Moncrief says, "We talked with the mother. We told her that it was experimental: *it had never been done* before we *had never done* it before."

Notice the use of the past perfect tense in the sentence above. We use this tense to contrast two actions that took place in the past, one action having taken place before the other. In this case, Dr. Moncrief and his medical team told Anthony's mother (one action in the past) about an experimental procedure that had never taken place before (prior to the time of their talk). The time line below may make the explanation clearer:

The procedure had never been used. Moncrief talked to Mrs.Austin

before 1981 1981 Now

Now complete the sentences below with the appropriate form of the past perfect. The first one has been done for you.

1. (have) Before Anthony rappelled down the 90-foot cliff, he _had_ never _had_ such an experience before.
2. (lose) When Anthony arrived at the airport, he realized that the airline ______________ his luggage.
3. (try) The children at Moncrief's camp were encouraged to do things they _____ never _____ before.
4. (get) Sathya felt a lot better after the doctor ________________ most of the fluid out of him.
5. (do) Sathya rappelled down the 90-foot cliff just as he ________________ the previous summer.
6. (go) Before Anthony went to Camp Courage, he _____ never _____ away from home alone before.

VOCABULARY IN CONTEXT

Based on the context clues in the sentences below, guess the meaning of the *italicized* words. First circle the words that helped you guess the meaning. Then write a definition in the space provided.

1. The new *dialysis* technique used a simple plastic bag and tubing connected to the abdominal cavity to cleanse the system of impurities.

   dialysis: ____

2. We told her that it was *experimental*. It had never been done before; we had never done it before.

   experimental: ____

3. The camp *infirmary* is well stocked with all the medicines he needs.

   infirmary: ____

4. Most of them have had the problem of facing death, and that probably has the single most important impact on them. It has made them recognize their *mortality*.

   mortality: ____

5. You've had to face things a lot scarier than this, Shanika. You've had a liver transplant. You've had to go *under the knife*. You've had to go into surgery.

   under the knife: ______________________________

6. Whether you go down this cliff or don't go down this cliff is not important. But recognizing that you have a fear of it and dealing with that is important. *Coming to grips* with fear and pain is something that happens a lot here at the Moncrief Ranch.

   coming to grips: ______________________________

Postviewing

DO YOU AGREE OR DISAGREE?

Work with a small group of students and explain why you agree or disagree with the following statements. Choose one student in the group to be the leader. Later, the group leaders can report the opinions of their group members to the rest of the class.

1. Children with chronic medical problems should not be segregated from "healthy" children. An effort should be made to integrate all types of children at a summer camp.
2. The counselors at Camp Courage push the children too much to perform difficult tasks. Children who are seriously ill should not be pushed so much.
3. People's characters develop more through positive experiences than through negative ones.
4. It is only because of advances in medicine and medical technology that places like Camp Courage can exist.
5. It's more important for society to use its financial resources to promote the development of healthy children, not sick ones.
6. In achieving goals, people's attitudes are more important than their abilities.

RELATED READING

Read the passage below to answer the questions that follow.

A Medical Miracle: Heart Transplants

On December 3, 1967, Dr. Christian Barnard, a South African surgeon performed an operation that had been considered unthinkable: he took a human heart from one body and successfully transplanted it into another. The heart was taken from a fatally injured accident victim and put into the body of Louis

Washansky, a South African grocer with an incurable heart ailment. Eighteen days after the surgery, Washansky died of double pneumonia, a disease that resulted from the suppression of his immune system by the drugs that had been administered to repress the body's mechanism of "rejection."

The mechanism of rejection is set in motion when foreign cells are introduced into the biological system, which then responds by sending "attack cells" to destroy the foreign ones. The "rejection" mechanism is one of the major obstacles to successful transplant surgery. But since the development of a drug called "immunosuppressant cyclosporine" doctors have been able to handle the rejection mechanism. By the late 1970s, each year hundreds of transplant operations were being performed, which extended the lives of patients with otherwise incurable heart diseases.

Of all the organ transplant operations, heart transplants are the most difficult to perform. The heart is more sensitive to lack of blood supply than the other organs. It must therefore be removed from the donor's body quickly and used for transplant surgery within a brief period since it can be preserved only for a short time. Artificial hearts have also been developed for transplant surgery; however, they are often not feasible because they require a cumbersome external power supply.

Source: *Encyclopedia Britannica,* Vol. 12. Encyclopedia Britannica, Inc., 1992.

QUESTIONS

Scan the reading passage for the answers to the following questions. When you locate the portion of the reading with the answer to a particular question, underline it and write the question's number next to it to justify your answers when you review them with the class. Finally, write full-sentence answers to the questions without looking back at the text.

1. What happened on December 3, 1967?

____

____

2. Who were the heart donor and the heart recipient?

____

____

3. Was the operation a success?

____

____

4. How long did the patient live after the transplant surgery?

____

____

5. What has been a major obstacle to successful heart transplant surgery?

____

____

6. What development allowed doctors to handle this obstacle better?

____

____

7. Why is heart transplant surgery particularly difficult to perform?

____

____

8. Have artificial hearts been a feasible alternative to human hearts in transplant surgery?

____

____

NOTABLE QUOTES

Write a paragraph explaining and commenting on the statement below.

"If you have rappelled down a cliff once, you have rappelled down a cliff for the rest of your life."

— Dr. Jack Moncrief

PROJECTS

IDEA #1

Write a letter to Dr. Moncrief in which you tell him your reaction to watching the program about Camp Courage. What aspects of the camp's program particularly impressed you? What aspects do you have questions about? Be sure to incorporate into your letter at least five or six of the vocabulary words studied in this chapter.

OR

Do some research on special programs and facilities available to handicapped children in your country. Write a short paper about your findings.

IDEA #2

Go to the library and do some research on a particular medical advance that has allowed people with a particular disease to live longer and perhaps more fulfilling lives. Describe the medical advance in as much detail as you can and explain how it has affected the lives of people who have the particular illness it addresses.

SUGGESTED READINGS

Batshaw, Mark, and Perret, Yvonne. *Children with Handicaps: A Medical Primer.* Brookes, 1986.

Burris, Dan and Thomsen, Patti. *Medical Advances: A User Friendly Guide to the Latest Technology*. International Management Publications, 1992.

Frist, William H. *Transplant: A Heart Surgeon's Account of the Life & Death Drama of the New Medicine*. Fawcett, 1990.

Goldfarb, Lori, et.al. *Meeting the Challenge of Disability or Chronic Illness*, Brookes, 1986.

Segment 7

The Power of Mimi

from *20/20*,12/29/89
Runtime: 15:58
Begin: 01:08:45

Mimi Silbert has helped former criminals make the transition from ex-convict to useful member of society.

Previewing

KEY QUESTIONS

1 Do prisons do a good job of reforming criminal behavior? Why or why not?
2. After a person has served time in prison, is it hard for him or her to become integrated back into society? Why or why not?
3. Is it possible for drug addicts and criminals to be successfully treated and reintegrated into society?

DISCUSSION QUESTIONS

1. Should government money be spent to treat criminals and drug addicts?

____

____

2. What do you think would be the most important features of a program that helped criminals and drug addicts successfully integrate into society?

____

____

PREDICTION

Based on the title of the segment, *The Power of Mimi*, the discussion you've had with your classmates, and your own knowledge, what information do you think will be included in this video segment?

1. ______________________________
2. ______________________________
3. ______________________________
4. ______________________________

ESSENTIAL WORDS TO KNOW

The following words are used in the video segment. Familiarize yourself with the meanings of the words before you view the segment. First a general definition is given. In the space that follows, write a sentence that includes the *italicized* word.

1. *bleeding-heart liberal* = a person who some people consider to be overly sympathetic to the poor and disadvantaged

   ____

2. *rehabilitate* = to change a behavior considered psychologically and/or physically unhealthy into one that is healthy

   ____

3 *hostile* = angry, full of resentment

   ____

4 *role model* = a person who sets a good example to those around him or her

   ____

5. *ex-convict* = someone who was once in prison

   ____

Global Viewing

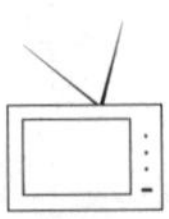

CONFIRMING YOUR PREDICTIONS

Watch the entire video segment. As you watch, put a check (✓) on page 75 next to any of your predictions that were mentioned.

01:08:45-
01:24:43

Prediction 1 ______
Prediction 2 ______
Prediction 3 ______
Prediction 4 ______

MAJOR POINTS/MAIN IDEA

01:08:45-
01:24:43

Work with a partner and write down some of the major points made in the program.

____

____

____

____

____

____

Compare your list with those of other students in the class. Have you written down the same points or different ones? Of all the various points, which do you consider to be the most important? Write it below.

____

____

THINK, WRITE, AND DISCUSS

01:08:45-
01:24:43

Watch the entire segment and answer the questions below. If you wish, you may take notes on a separate sheet of paper. After you've finished, discuss your answers with the class.

1. Why has Delancey Street, Mimi Silbert's rehabilitation program for ex-convicts, been successful?

   ____

   ____

2. What kind of person is Mimi? What adjectives would you use to describe her?

   ____

   ____

3. Did the ex-convicts expect to become rehabilitated when they first entered Delancey Street? Why or why not?

4. How is Delancey Street financed?

5. Based on what you have seen in the segment, what is the best way to help people overcome their past and change for the better?

Intensive Viewing

LISTENING CLOZE

01:09:30-
01:09:58

Watch this part of the segment and fill in the missing words. Then compare your answers with those of other students.

Hugh Downs: If a thousand __________ of light will solve America's social problems, we need to __________ Mimi Silbert. Now, she is a __________ with an approach that's as special as her own personality. She takes ex-convicts, ______________ losers from the __________ of the heap, and transforms them into responsible, well-paid members of the community. The next time somebody's talking about how our criminal population is __________, and that we're out of ______________, you can tell them about "The Power of Mimi."

NOTETAKING

01:10:23-
01:15:40

The notetaking form on page 77 has been partially filled in for you. First read through the notes. Then view the videotape, filling in the missing information. When you finish, compare your notes with those of others to check your comprehension.

1. The Justice Department is interested in Delancey Street because:

   a. ____________________

   b. ____________________

2. Louise Likerman

   a. Before Delancey Street, she was: ____________________

      1. ____________

      2. ____________

   b. Her job at Delancey Street is: ____________________

   ____

3. Each resident must learn: ____________________

4. When they first come to Delancey Street residents are:

   a. vicious

   b. ____________

   c. ____________

5. Mimi Silbert's point is: ____________________

   ____

6. The resident's behavior

   a. Same as prison because residents enforce their own systems of behavior

   b. Different because they are enforcing systems of ____________

   ____

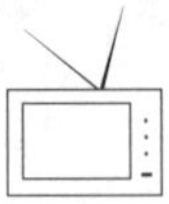

BUILDING DELANCEY STREET

01:18:06-01:21:51

In this part of the segment, Bob Brown describes one of Mimi Silbert's "most extraordinary and improbable dreams of change for Delancey Street." What are some of the things you learn about her dream? Take notes in the space below, then compare your notes to those of another student.

____

____

____

____

Language Focus

COLLOQUIAL ENGLISH: IDIOMS

Work with a partner and read the following part of the segment. Then match the *italicized* words with their meanings. Compare your answers with those of other students in the class.

Ms. Silbert: We have people coming in who have really committed horrible crimes. What matters is that you're willing to say, "I take responsibility. I did it. I *screwed up* my life. I hurt other people, and I want to change, and please help me." And that's really what our whole interview is. We sit down, we ask you to tell your story. Usually you tell it the way you talk to most social workers, which is you're a victim. "Poor me. I didn't have a mommy, I didn't have a daddy. Everything is wrong . . ."

Brown: They *know the drill*, don't they?

Ms. Silbert: They *have it down*. They know when to cry. "I'm very hostile. I don't want to be hostile no more."* You know, and then we say, "*Cut the crap*."** You see, most of the time, they just say it to get by.

*Here, Ms. Silbert is offering some typical quotes from the convicts, some of whom speak non-standard English; the use of the double negative in "I don't want to be hostile no more" is non-standard. In standard English, the sentence would be, "I don't want to be hostile any more."

**this expression is quite colloquial and should be used only in the most informal conversation

______	1. screwed up	a. stop talking in a dishonest way
______	2. know the drill	b. failed
______	3. have it down	c. know (something) very well
______	4. cut the crap	d. know the expected thing to say

VOCABULARY IN CONTEXT

Guess the meanings of the *italicized* words in the sentences below. Circle the context clues, those words in each sentence that help you to guess the meaning. Then write your definition in the space provided.

1. I did anything and everything for money, and I really didn't think I'd ever actually *clean up* and become like a regular kind of person that didn't use drugs.

   clean up: ____

2. I have never *deluded* myself that people coming into Delancey Street are nice. They are not nice.

   deluded: ____

3. We try to make our whole lives a neighborhood in which some of the older people who really are never going to be hotshots in life can be wise and can really sit and give experience . . . So there's a *tremendous support system*.

   tremendous support system: ______________________________

4. When I finally *put my mind to it*, I went and studied for a month or so and went up to take the test.

   put my mind to it: ____

Postviewing

DO YOU AGREE OR DISAGREE?

Work with a small group of students and discuss whether you agree or disagree with the following statements. Choose one student to be the group leader. Following your small-group discussion, the leader can share the various opinions of the group members with the rest of the class.

1. Ex-convicts don't deserve to be treated as well as they are at Delancey Street.
2. Delancey Street works because it offers criminals social rehabilitation rather than punishment.

3. It's very important that the residents of Delancey Street make and enforce their own rules.
4. Human beings need to feel that they are part of a strong community in order to be happy and productive.
5. An organization like Delancey Street would probably be successful in my country too.
6. Delancey Street works mainly because of Mimi Silbert's very special, charismatic personality; if another person tried to manage the organization, it would not be so successful.

RELATED READING 1

Read the passage below to answer for the questions that follow.

The Prison Question

For many years sociologists, psychologists, penologists, and the general public have become increasingly aware of the ineffectiveness of the prison system in the United States. Rarely are convicted criminals rehabilitated after serving time in prison; in fact, they are often far more capable of committing crime after leaving prison than they were upon first entering. This is because the "prison world" becomes a kind of ghetto into which all the worst social problems of urban life are compressed: overcrowding, poverty, violence, despair. Naturally there is very little hope of social rehabilitation under such conditions.

To make matters worse, prisons are short of professional staff; trained, qualified psychologists and criminologists who might offer much-needed treatment to the inmates are often reluctant to work in a prison environment. According to statistics, only 10 percent of prison staff deal with treatment of inmates, while 60 percent deal with their security and custody.

Some alternative approaches to the traditional prison system, such as performing community service, paying fines, or paying back the victims of crime, have been developed and implemented on a limited scale in the United States. Such alternative systems are currently in use on a much grander scale in some European countries, particularly those in Scandinavia.

COMPREHENSION QUESTIONS

Scan the above text to find the answers to the questions on page 81. When you have located the part of the text with the answer to a question, underline it and write the number next to the underlined words so that you can justify your answer later. Then write short answers to the questions on page 81 in your own words.

1. Do social scientists have a positive feeling about the U.S. prison system?

____

____

2 Why are convicted criminals often more capable of committing crime after leaving prison than before entering?

____

____

3. What percentage of the prison staff deals with the treating inmates?

____

____

4. What are some alternative approaches to the traditional prison system?

____

____

RELATED READING 2

Read the passage below to prepare for the questions that follow.

The Power of Positive Thinking at Delancey Street

In the fall of 1991, Mimi Silbert, founder of Delancey Street, the rehabilitation center for ex-convicts, was one of six people to receive an America's Award from the Positive Thinking Foundation. The Foundation was co-founded by Dr. Norman Vincent Peale, whose book, *The Power of Positive Thinking*, first published in 1953, was a world-wide best-seller. Says Dr. Peale of the award recipients, "These unsung heroes personify the American character and spirit. They are ordinary people who are extraordinary examples of values that make our country great."

Silbert's center, Delancey Street, was named after a major street on New York City's lower east side, where immigrants from many different nations lived around the turn of the century. Silbert, who is herself from an immigrant community in Boston, explains that Delancey Street incorporates a lot of the same values that helped inhabitants of early immigrant communities rise to success. As Silbert says of her program at Delancey Street, "I've duplicated here what worked for me in that neighborhood where everybody looked out for everybody else as we struggled upward. It was like holding hands while climbing a mountain. Together we rise or together we fall."

As of 1992, five hundred residents lived at the Delancey Street complex in San Francisco, California. An additional five hundred participants are going through Delancey Street programs in other parts of the United States: Brewster, New

York: Greensboro, North Carolina: and San Juan Pueblo, New Mexico. The Delancey Street programs are almost completely self-supporting, and their net income is about three million dollars per year. Silbert's next plan is to develop a training institute that would teach the staff of other rehabilitation centers the techniques that work so successfully at Delancey Street. The Delancey Street program could become a national model for ex-prisoner rehabilitation.

As Mimi Silbert says, in the typical "positive thinking" style that earned her an America's Award, "The whole point of Delancey Street is that ordinary people, as a matter of fact less-than-ordinary people — our people are truly the losers of society — can, if they pool their resources, if they're structured and disciplined and taught, and then motivated and believed in, can do extraordinary things."

Source: "Hitting the Bottom Can Be the Beginning." Parade Magazine. March 15, 1992.

QUESTIONS

Write questions about the above reading passage beginning each with the words below. Make sure that the form of your questions is correct. Then, ask your partner to answer the questions without looking back at the text.

1. Who ______________________________?
2. When ______________________________?
3. What ______________________________?
4. Why ______________________________?
5. How many ______________________________?
6. How much ______________________________?

NOTABLE QUOTES

Write your reaction to one of the following quotations. Explain what the statement means and agree or disagree with it.

"There is no such thing as going through life okay. You go along, you trip and you fall. At that moment, you have to gather your own courage up and you have to say, 'Okay, I did it wrong, I messed up once again. I wish I didn't, but damn it, here I come!'"

—Mimi Silbert

"But when I think about it, truly think about it, it's scary. I'm changing into somebody that I do not know and changing into somebody that four or five years ago I thought was impossible to do."

—Allen Loucks

PROJECT

IDEA #1

Make a radio program. With the members of your group, discuss what you consider to be the most important and necessary features of a successful rehabilitation program for prisoners and ex-convicts. Talk about why you feel many present-day rehabilitation programs are unsuccessful. Write selected student comments in your radio script. Each of the written comments should contain at least two of the new vocabulary words or expressions learned in the course of studying the program *The Power of Mimi.* In contributing lines to the radio tapescript, students should not repeat new words already used by other students. After the tapescript is completed and checked by your teacher, you may make your radio program. Each student can record his or her comments on the tape. Choose one member of your group to be the radio program moderator. The moderator will introduce each of the students and give an introduction and a conclusion to the program.

Plan to do a rehearsal and a final program. Evaluate your rehearsal and decide what needs to be changed in order to make an improved final program (pronunciation, grammar, choice of words, etc.). After completing the final program, each group will play its program for the rest of the class. After listening to all of the programs, the class can hold a contest to choose the best program.

IDEA #2

Go to the library and do some research on the subjects of criminal rehabilitation and prison reform in your country or in other countries. To aid yourself in this research, you may wish to consult the suggested reading list below. The short passage included in this chapter may provide an introduction to the topic.

SUGGESTED READINGS

Atkins, B. and Glick, H., editors. *Prisons, Protest, and Politics*. Prentice-Hall, 1972.

Blom-Cooper, Louis, editor. *Progress in Penal Reform*. Oxford, 1975.

Ohlin, Lloyd E. editor. *Prisoners in America: Perspectives on Our Correctional System*. Prentice-Hall, 1973.

Peale, Norman Vincent. *The Power of Positive Thinking: Special 35th Anniversary Edition*. Prentice-Hall, 1987.

Vagg, John and Maguire, Mike, editors. *Accountability and Prisons: Opening Up a Closed World*. Methuen, 1986.

Segment 8

Annie DeMartino

from *World News Tonight,* 11/24/89
Runtime: 4:50
Begin: 01:24:51

Annie DeMartino is a social worker and city council member who has helped hundreds of homeless families find affordable housing.

Previewing

KEY QUESTIONS

1. What are some of the causes of homelessness?
2. Should the government be required to help homeless people? Why or why not?
3. Why has there been so little success in solving the problem of homelessness?

DISCUSSION QUESTIONS

1. Are there now more or fewer homeless people than in the past? Why or why not?

   ____

   ____

2. Do you think homeless people can help themselves? If yes, how?

   ____

   ____

3. What do you think a government can do to help the homeless?

   ____

   ____

PREDICTION

Based on the title of the segment, *Annie DeMartino,* the discussion you've had with your classmates, and your own knowledge, what information do you think will be included in this video segment?

1. ______________________________
2. ______________________________
3. ______________________________
4. ______________________________

ESSENTIAL WORDS TO KNOW

The *italicized* words in the sentences below are used in the video. Read the sentences, then match the words with their meanings.

disgrace: It is a *disgrace* that people are hungry when the word has a huge food supply.

tenacious: Because he was *tenacious* and wouldn't take no for an answer, he convinced the foundation to fund his project.

orphanage: She grew up in an *orphanage* because her parents died in a car accident.

destitute: He was left *destitute,* with no food, clothing or a home.

welfare: The politician came up with a *welfare* program to help those who were too sick to work or couldn't find a job.

_____ 1. disgrace
_____ 2. tenacious
_____ 3. orphanage
_____ 4. destitute
_____ 5. welfare

a. persistent
b. lacking the necessities of life
c. shame, dishonor
d. an institution that cares for children without families
e. a financial support system funded by the government

Global Viewing

CONFIRMING YOUR PREDICTIONS

01:24:51-
01:29:41

Watch the entire video segment. As you watch, put a check (✓) next to any of your predictions that were mentioned.

Prediction 1 _____
Prediction 2 _____
Prediction 3 _____
Prediction 4 _____

MAJOR POINTS/MAIN IDEA

Write down some of the major points made in the program.

01:24:51-01:29:41

Now work with a partner and decide which is the most important and write it below in a complete sentence.

THINK, WRITE, AND DISCUSS

Watch the entire segment and answer the questions below. If you wish, you may take notes on a separate sheet of paper. After you've finished, discuss your answers with the class.

01:24:51-01:29:41

1. What kind of work does Annie DeMartino do?

2. Why is Annie DeMartino so committed to her work?

3. How has she succeeded in finding homes for so many homeless people? Why has she been able to succeed where others have failed?

4. Why does Annie say that she works so hard?

Intensive Viewing

LISTENING CLOZE

Watch this part of the segment and fill in the missing words. Then compare your answers with those of other students.

01:24:58-
01:25:16

Sam Donaldson: And finally tonight, our Person of the Week, whose ____________________ to the Thanksgiving season _________ all year round. She first __________ our attention in 1987 when she started a program which found homes for every _____________ family in her town. And when we ____________ back, we discovered that she's still ____ it.

MAKING THE COMMUNITY BETTER

Watch this part of the segment and answer the questions that follow.

01:25:54-
01:28:08

1. What is Annie's role as Fitchburg's newest council member?

   ____

   ____

2. How did Annie help Tracy?

   ____

   ____

3. How did Annie convince landlords to rent Tracy and other homeless people apartments?

   ____

   ____

4. What was Annie's life like when she first came to the United States?

   ____

   ____

5. What does she say about recovering from poverty?

   ____

   ____

6. How is Annie helping housing developers?

   ____

   ____

7. How does this also help low-income families?

____

____

8. Describe Fitchburg, the community Annie lives in.

____

____

Language Focus

WORD FORMS

The following words were used in the segment. Complete the grid below by adding the other forms of the given words:

NOUN	VERB	ADJECTIVE
1. contribution	____________	____________
2. ____________	to believe	____________
3. ____________	____________	helpful
4. ____________	to elect	____________
5. inspiration	____________	____________

PHRASAL VERBS

The following sentences are from the segment. Fill in the blanks with the correct form of the phrasal verb.

grow up push through work for end up

1. She ____________ in an orphanage in Ireland.
2. Annie ______________ a single mother on welfare.
3. In a few hours, Annie ______________________ the paperwork for Andy's Medicaid.
4. She is everything that we ____________ everyday, and she's in college.

JOINING SENTENCES WITH *WHEN*

When joins two sentences describing events that happened at the same time. Look at the example from the video.

When we checked back, we discovered she's still at it.

The example combines the following two sentences.

We checked back.

We discovered she's still at it.

We could also change the order of the combined sentences, as seen below. In this case, no comma is used to separate the two clauses as was done in the first example.

We discovered she's still at it *when* we checked back.

Join the following sentences using *when*. You may put the clause using *when* at the beginning or at the end of your new sentence. Be sure to use a comma if needed.

1. Mr. Martin called his wife.
   He needed her advice.

   ____

   ____

2. She came into the apartment.
   The phone rang.

   ____

   ____

3. Ms. Anderson bought a new car.
   She learned how to drive.

   ____

   ____

4. I sang and danced.
   The band played my favorite song.

   ____

   ____

5. They heard that their friends were visiting.
   They cleaned their apartment.

   ____

   ____

JOINING SENTENCES WITH *WHERE*

Where joins two sentences that refer to the same place. Look at the example from the video.

> Annie is still working at the Department of Social Services, *where* we found her two years ago.

This example joins the following two sentences.

> Annie is still working at the Department of Social Services.
>
> We found her there two years ago.

Notice that *there* is used to refer to the *Department of Social Services*. We delete *there* when we join the two sentences with *where*. Also note that the order of the clauses cannot be changed; we must first specify the place and then refer back to it using *where*. We also use a comma to separate the two clauses.

Join the following sentences using *where*.

1. Anna went to art school.
   She studied design there.

   ____

   ____

2. They all went to the store.
   Stephanie bought her new sweater there the week before.

   ____

   ____

3. Hassan went to the new-car showroom.
   He bought a new car there.

   ____

   ____

4. Andrew went to the theater.
   He played the lead character in a comedy there.

   ____

   ____

5. Michiko and Marie Laure tried to find the little café.
   They had lunch there several weeks ago.

   ____

   ____

Postviewing

DO YOU AGREE OR DISAGREE?

Form small groups and tell if you agree or disagree with the following statements.

1. Every citizen of every country has the right to a place to live — even if he or she is poor or unemployed.
2. Annie is right when she says it's a disgrace that there are homeless people in one of the richest countries in the world.
3. If people in America are homeless, it's probably their own fault.
4. There are homeless people in my country, too.
5. Annie is right when she says it's better for the working poor to have the option to own their own homes.
6. Annie DeMartino is right when she says she owes a debt to America.

RELATED READING 1

Read the article *Homeless in America* to answer the questions on page 93.

Homeless in America

The Coalition for the Homeless is an organization that seeks to address the needs of the homeless population in the United States. It is a network of agencies, some of which provide food and shelter to the homeless population, and some of which advocate the passage of legislation that would give every American the right to a place to call home. According to the Coalition's 1992 statistics, of the two hundred million inhabitants of the United States, up to three million are homeless — and the numbers are growing. Since the late 1970s, soaring real estate prices, large cuts in federally subsidized housing programs, and economic recession have made it impossible for many Americans to meet housing costs. Sadly, this has resulted in a number of persons being evicted from their homes and/or unable to find new affordable homes. According to a 1989 survey, families with children appear to be the fastest-growing segment of the homeless population, comprising 39% of it. The old stereotype of a homeless person, that of the single alcoholic male, no longer applies. A much broader section of the population now finds itself without permanent shelter. Even worse, once people become homeless, they often find it impossible to find a job, since most employers require their prospective employees to provide a residence address on a job application.

Source: *Coalition for the Homeless Fact Sheet,* 1992.

QUESTIONS

Scan the passage on page 92 for answers to the questions below. Underline those portions of the text in which the answer is contained. Then cover the text, and write answers to the questions.

1. What kind of an organization is the Coalition for the Homeless?

   ____

   ____

2. How many people in the United States have no homes?

   ____

   ____

3. What are some reasons for the rising numbers of homeless people?

   ____

   ____

4. What is the fastest growing segment of the homeless population?

   ____

   ____

5. Why do homeless people often have problems finding jobs?

   ____

   ____

RELATED READING 2

Read the article *Habitat for Humanity* to prepare for the task that follows.

Habitat for Humanity

Habitat for Humanity is a non-profit, volunteer organization, supported and promoted by former U.S. President Jimmy Carter. The organization works in partnership with low-income families and individuals to build decent, affordable housing for them. Habitat acquires buildings or land and then trains eligible low-income people in construction so that they can help build their own homes together with hundreds of volunteers who work at the building site.

Apartments completed by Habitat volunteers are then sold to homesteaders or the eligible homeless persons at Habitat's cost only. Each homesteader is given a thirty-year interest-free mortgage.

Habitat for Humanity has more than six hundred affiliated projects in communities across the United States, as well as more than one hundred sponsored projects in developing countries. Funding for Habitat comes from individual donors, churches, corporations, foundations, and other organizations; Habitat receives no government contributions.

Source: *Habitat for Humanity Newsletter*, Spring 1992.

QUESTIONS

Write questions about the preceding passage by completing the lines below. Then ask your partner to answer the questions without looking back at the reading passage.

1. What ____?
2. Who ____?
3. How ____?
4. How many ____?
5. Where ____?

NOTABLE QUOTES

Write a paper or prepare an oral report in which you give your reaction to one of the quotations below. Explain what Annie DeMartino is saying and agree or disagree with her. Use four or five of the new words and expressions you learned in the segment.

"I see broken promises every day in children's eyes and it's very sad—and—that's why I work so hard."

— Annie DeMartino

"Change starts at the top, not at the bottom. And if I've got to do something different, I'd better learn how the city government works."

— Annie DeMartino

PROJECT

IDEA #1

Write a letter to Annie DeMartino, congratulating her on all the good work she has done. Mention some of her accomplishments. Use four or five of the words and expressions you learned in the segment.

IDEA #2

You may wish to do further research on the problem of homelessness in the United States and/or in other countries of the world. Write to one of the organizations below for information about world homelessness.

United Nations
Department of Public Information
New York, New York 10017
U.S.A.

OR

The Coalition for the Homeless
1621 Connecticut Ave. NW, 4th floor
Washington, D.C. 20009
U.S.A.

SUGGESTED READINGS

Kozol, Jonathan. *Rachel and Her Children: Homeless Families in America.* Crown, 1988.

Porter, Kathryn H., *Poverty in Rural America: A National Overview.* Washington, D.C.: Center on Budget and Policy Priorities, 1989.

Responding to the Needs of the Homeless and Hungry. American Red Cross, 1989.

Schorr, Lisbeth B., with Daniel Schorr. *Within Our Reach*. Doubleday, 1988.

What Corporations Can Do To End Homelessness. Washington, D.C.: The National Alliance to End Homelessness, 1990.

Working with Homeless People. Columbia University Community Services, 1988.

Segment 9

James Grant

from *World News Tonight,* 9/28/90
Runtime: 5:19
Begin: 01:29:49

James Grant is a man with a dream of helping the world's children—and enough energy and imagination to make it happen.

Previewing

KEY QUESTIONS

1. In how many nations of the world do children suffer from lack of sufficient nutrition and medical treatment?
2. What international organizations work toward relieving the suffering of the world's disadvantaged children?
3. Do world leaders give enough attention to the problems of children in poor, developing nations?

DISCUSSION QUESTIONS

1. What things have been done to help poverty-stricken children?

____

____

2. Do you think this is enough? Why or why not?

____

____

3. Who should be responsibile for improving the living conditions of the world's poverty-stricken children? Why?

____

____

PREDICTION

Based on the title of the segment, *James Grant*, the discussion you've had with your classmates, and your own knowledge, what information do you think will be included in this video segment?

1. ______________________________

2. ______________________________

3. ______________________________

ESSENTIAL WORDS TO KNOW

The *italicized* words in the sentences below are used in the video. Read the sentences, then match the words with their meanings.

vaccine:	The doctor gave the child a *vaccine* so that she wouldn't get the flu.
rehydrated:	He had to be *rehydrated* because he had been deprived of water.
proselytizer:	James Grant is described as a *proselytizer* because he is always trying to persuade people to accept his point of view.
optimist:	Since she is an *optimist*, she always sees the future positively.

_____	1. vaccine	a. a person who seeks to convert others to his or her beliefs.
_____	2. rehydrate	b. an injection that prevents someone from getting a certain disease.
_____	3. proselytizer	c. someone who sees some good in everything.
_____	4. optimist	d. to give back water to someone who has been without water.

Global Viewing

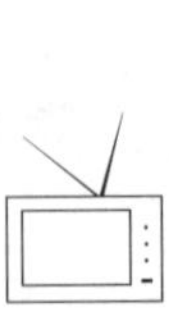

01:29:49-
01:35:08

CONFIRMING YOUR PREDICTIONS

Watch the entire video segment. As you watch, put a check (✓) next to any of your predictions that were mentioned.

Prediction 1 ______

Prediction 2 ______

Prediction 3 ______

MAJOR POINTS/MAIN IDEA

Work with a partner and write down a few of the major points made in the program.

01:29:49-
01:35:08

THINK, WRITE, AND DISCUSS

Watch the entire segment and answer the questions below. If you wish, you may take notes on a separate sheet of paper. After you've finished, discuss your answers with the class.

01:29:49-
01:35:08

1. Do children in all parts of the world have adequate access to medical treatment?

2. What are some of the leading causes of death among children? Is it possible to prevent any of these deaths?

3. What sort of life and work experience did James Grant have before coming to UNICEF?

4. How did James Grant manage to arrange the World Summit for Children?

Intensive Viewing

LISTENING CLOZE

Watch this part of the segment and fill in the missing words. Then compare
01:29:54- your answers with those of another student.
01:30:25

Peter Jennings: Finally this evening, our Person of the Week. We have chosen a man with a ______________ to save lives, to save many lives. This weekend, when more than 70 of the world's ______________ sit down together here in New York, it will be the first time they have been brought together by the ____________ of the world's least __________________ children. A world ____________ for children because 40,000 children die today and die every day. Tonight we choose a man who insisted that this summit be held and __________________ until it was.

READ AND LOOK UP: PRACTICE NARRATING

Turn off the sound and closed-captioning as you watch the beginning part
01:29:54- of the segment once again. This time you and other students in the class
01:30:25 will take turns reading the program narration above. When reading the narration, use the "read and look up" method. Using this method, don't just read words on a page. Rather, look at the words in a passage, phrase by phrase, and then look up and speak each phrase. This method will help capture the natural grouping of words in spoken English.

Mark the different phrases in the passage with a slash (/).

Example:

> Ten years ago / Grant was considered / the natural choice / to head UNICEF. / There is / no pocket of poverty / anywhere in the world, /no place / where there is disease and starvation / with which / he is unfamiliar.

UNAVAILABLE CURES

James Grant and Peter Jennings mention three cures that are cheap and
01:30:26- simple to administer, yet unavailable to tens of thousands of children. List
01:31:30 the cures below.

1. ____
2. ____
3. ____

HOW MUCH MONEY?

In this part of the segment, Peter Jennings discusses the amount of money needed to save the lives of 50,000 children each year. What is this amount of money, and what does Grant say about it in regard to American tobacco companies?

01:31:30-
01:31:47

JAMES GRANT'S BACKGROUND

This portion of the program tells some facts about James Grant's early life and career. As you listen, write as many points as you can. Later, compare your points with those of another student. Are they the same or different?

01:32:41-
01:33:21

Language Focus

STRUCTURE PRACTICE: ADJECTIVE CLAUSES

Combine the sentences below by using adjective clauses. Look at the following example.

James Grant is a man.

He is 68 years old.

James Grant is a man *who* is 68 years old.

1. James Grant believes we have the power to save a great number of children.

   These children die every year from readily preventable causes.

   ---

2. Grant is the Unicef Executive Director.

   He insisted that the world leaders gather together for a summit conference on the plight of the world's least fortunate children.

   ____

   ____

3. The packet contains salt, potassium, and glucose.

   You mix it with a liter of water to make a cure.

   ____

   ____

4. The World Summit for Children was an international conference.

   It was attended by seventy heads of state.

   ____

   ____

5. The World Summit for Children was an extremely important meeting of world leaders.

   It was held in New York City.

   ____

   ____

THE SUBJUNCTIVE

In his introduction of James Grant, Peter Jennings says:

> Tonight we meet the man who *insisted that this summit be held* and persisted until it was.

Notice that the verb *be* following the word *summit* is not conjugated; rather, the verb is used in its base form. If we wish to make the verb negative, we simply place the word *not* before it. Below are some verbs with which the subjunctive is commonly used.

insist recommend suggest

The subjunctive is also used with expressions like:

It is [important / essential / crucial / imperative] that . . .

Using some of the verbs and expressionson page 102, write some sentences about the program. Look at the examples below.

James Grant feels it is crucial that parents be taught to make a sugar/salt solution to cure their children of dehydration.

It is imperative that the suffering of the world's least fortunate children not be ignored.

1. ____

2. ____

3. ____

Postviewing

DO YOU AGREE OR DISAGREE?

Work with a small group of students and discuss whether you agree or disagree with the following statements. Choose one of the members of your group to be the leader. Later the leader can share the opinions of your group with the other members of the class.

1. There should be a world summit conference for children every year.
2. World leaders generally give priority to the problems of the world's children.
3. For many people, the world's least fortunate children are so far away that their problems seem unreal.
4. It is the responsibility of the governments of developed countries to give aid to help children in developing countries.
5. A society's level of humanity can be measured by how it treats its children.

RELATED READING

Read the article *The United Nation's Children's Fund: UNICEF* and answer the questions that follow.

The United Nation's Children's Fund: UNICEF

The United Nations Children's Fund, or UNICEF, was founded on December 11, 1946, as a semi-autonomous agency of the United Nations. UNICEF was first named the United Nation's International Children's Emergency Fund because it was originally created to provide emergency food, clothing, and medical supplies to the children of war-torn Europe.

After the state of emergency in Europe had passed, UNICEF began long-range programs to fight disease, malnutrition, and illiteracy. Today, UNICEF's main task is to assist the governments of developing countries in improving the quality of life for almost a billion children. In over one hundred countries, UNICEF's programs promote maternal and child health care, day care, applied nutrition, sanitation, education, and social welfare services.

In 1965 UNICEF was awarded the Nobel Peace Prize for its role in promoting international brotherhood by bringing the nations of the world closer together through its variety of educational programs.

Although the main headquarters of UNICEF is in New York City, the organization has forty field offices around the world in such cities as Bangkok, Cairo, and Vienna.

Source: *Encyclopedia Britannica*, Vol. 12. Grolier, Inc., 1992.

QUESTIONS

Write short answers to the questions below. Be prepared to justify your answers by underlining the part of the text where you found each one. Then ask your partner to answer each of the questions without looking at his written answers or at the text.

1. When was UNICEF started?

____

____

2. What is its status regarding its relation to the United Nations?

____

____

3. What was the original purpose of UNICEF?

____

____

4. What is the main purpose of UNICEF today?

____

____

5. What happened in 1965?

____

____

NOTABLE QUOTES

Write a two-or three-paragraph paper or give a two- or three-minute talk explaining what the statement below means and agreeing or disagreeing with it.

"If we can do something to prevent the torment of these children, but don't — then we join the tormentors."

— James Grant

PROJECT

IDEA #1

Find out more information about UNICEF by writing to the organization's headquarters in New York City. The address is:

UNICEF
UNICEF House
3 United Nations Plaza
New York, New York 10017
United States of America

IDEA # 2

Below is a reprint of one of the most famous charters of the United Nations: *The Declaration of the Rights of the Child.* Choose one part of the declaration and write a paragraph explaining why you think it is particularly important.

The Declaration of the Rights of the Child

Whereas mankind owes to the child the best it has to give,

Now, therefore,

The General Assembly

Proclaims this Declaration of the Rights of the Child to the end that he may have a happy childhood and enjoy for his own good and for the good of society the rights and freedoms herein set forth, and calls upon parents, upon men and women as individuals and upon voluntary organizations, local authorities and national governments to recognize these rights and strive for their observance by legislative and other measures progressively taken in accordance with the following principles:

PRINCIPLE 1

The child shall enjoy all the rights set forth in this Declaration. All children, without any exception whatsoever, shall be entitled to these rights, without distinction or discrimination on account of race, color, sex, language, religion, political or other opinion, national or social origin, property, birth or other status, whether himself or of his family.

PRINCIPLE 2

The child can enjoy special protection, and shall be given opportunities and facilities, by law and by other means, to enable him to develop physically, mentally, morally, spiritually and socially in a healthy and normal manner and in conditions of freedom and dignity. In the enactment of laws for this purpose the best interests of the child shall be the paramount consideration.

PRINCIPLE 3

The child shall be entitled from his birth to a name and nationality.

PRINCIPLE 4

The child shall enjoy the benefits of social security. He shall be entitled to grow and develop in health; to this end special care and protection shall be provided both to him and to his mother, including adequate pre-natal and post-natal care. The child shall have the right to adequate nutrition, housing, recreation and medical services.

PRINCIPLE 5

The child who is physically, mentally or socially handicapped shall be given the special treatment, education and care required by his particular condition.

PRINCIPLE 6

The child, for the full and harmonious development of his personality, needs love and understanding. He shall, wherever possible, grow up in the care and under the responsibility of his parents, and in any case in an atmosphere of affection and of moral and material security; a child of tender years shall not, save in exceptional circumstances, be separated from his mother. Society and the public authorities shall have the duty to extend particular care to children without a family and to those without adequate means of support. Payment of state and other assistance toward the maintenance of children of large families is desirable.

PRINCIPLE 7

The child is entitled to receive education, which shall be free and compulsory, at least in the elementary stages. He shall be given an education which will promote his general culture, and enable him on a basis of equal opportunity to develop his abilities, his individual judgement, and his sense of moral and social responsibility, and to become a useful member of society.

The best interests of the child shall be the guiding principle of those responsible for his education and guidance; that responsibility lies in the first place with his parents.

The child shall have full opportunity for play and recreation, which should be directed to the same purposes as education; society and the public authorities shall endeavor to promote the enjoyment of this right.

PRINCIPLE 8

The child shall in all circumstances be among the first to receive protection and relief.

PRINCIPLE 9

The child shall be protected against all forms of neglect, cruelty and exploitation. He shall not be the subject of traffic, in any form. The child shall not be admitted to employment before an appropriate minimum age; he shall in no case be caused or permitted to engage in any occupation or employment which would prejudice his health or education, or interfere with his physical, mental or moral development.

PRINCIPLE 10

The child shall be protected from practices which may foster racial, religious and any other form of discrimination. He shall be brought up in a spirit of understanding, tolerance, friendship among peoples, peace and universal brotherhood and in full consciousness that his energy and talents should be devoted to the service of his fellow men.

SUGGESTED READINGS

Assessment of the Impact of Measures and Activities Accomplished During the International Youth Year. United Nations, 1987

A History of the United Nations, vol. I & II. St. Martin's, 1982, 1989.

The United Nations: How It Works, What It Does. St. Martin's, 1985.

World Summit for Children: UNICEF Information Kit. United Nations, New York, 29–30 September, 1990.

Segment 10
Andy Warhol

from *20/20*, 4/10/80
Runtime: 12:44
Begin: 01:35:16
Note: This segment is not closed-captioned.

Andy Warhol introduced pop art to America in the 1960s.

Previewing

KEY QUESTIONS

1. What is the most obvious difference between modern art and more traditional forms of art?
2. Can any subject be suitable as subject matter for art? Why or why not?
3. Should an artist's work be judged on its own, independent of the artist's personality, or can it be better understood if one knows something about the artist?

DISCUSSION

1. Do you prefer modern art or more traditional forms of art? Why?

___

___

2. Can you think of some world-famous artists known for their interesting life-styles or personalities?

___

___

3. How interested should the public be in the private life of an artist?

___

___

PREDICTION

Based on the title of the segment, *The Art of Andy Warhol*, the discussion you've had with your classmates, and your own knowledge, what information do you think will be included in this video segment?

1. ____________________

2. ____________________

3. ____________________

4. ____________________

ESSENTIAL WORDS TO KNOW

The *italicized* words will be used in the video. Read the words as they are used in the following sentences and try to come up with your own definition.

1. Some artists are criticized for being too *commercial* and focusing on money-making goals.

   commercial: ____________________

2. *Publicity* being one of Andy Warhol's strengths, he was always able to make himself noticed.

   publicity: ____________________

3. One of his most famous pieces of *pop art* is his painting of the Campbell's soup can.

   pop art: ____________________

4. Andy was an artist in many *media*, including painting, sculpture, writing, film, and television.

   media: ____________________

5. The artist's *promoter* traveled across the country trying to set up exhibits to show his client's art.

   promoter: ____________________

Global Viewing

CONFIRMING YOUR PREDICTIONS

01:35:16-
01:48:00

Watch the entire video segment. As you watch, put a check (✓) next to any of your predictions that were mentioned.

Prediction 1 ______

Prediction 2 ______

Prediction 3 ______

Prediction 4 ______

MAJOR POINTS/MAIN IDEA

With a partner, watch the entire segment and write down a few of the major points made in the program.

Compare the points you have chosen with those chosen by other students in the class. Does any one point seem most important or seem to include the others? Identify this main idea and write it below.

THINK, WRITE, AND DISCUSS

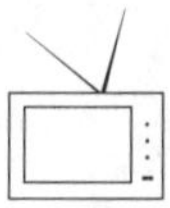

01:35:16-
01:48:00

Watch the entire segment and answer the questions below. If you wish, you may take notes on a separate sheet of paper. After you've finished, discuss your answers with the class.

1. What is the subject matter of Andy Warhol's art?

2. In what media besides painting did Andy Warhol work?

____

____

3. What are some of the different responses to Andy Warhol's art?

____

____

4. What was Andy Warhol's relationship to celebrities?

____

____

5. How do different people describe Andy Warhol and his artwork?

____

____

Intensive Viewing

LISTENING CLOZE

Watch this part of the segment and fill in the missing words. Then compare your answers with those of another student.

01:35:48-
01:36:25

Hugh Downs: Who is the ________ known and certainly the most controversial ___________ in America today? I'll give you some _________. He _____________ dollar bills, Coke _____________ , Marilyn Monroe, and this—a Campbell's ________ can. Andy Warhol, that's who. He makes art out of ________________ things, and uses many _________. You blink and he's doing something new. Lately, a major art opening and _____________________ of his new book of _____________________ — and now, just out, another ________ , *Popism,* about his life and _____________ in the 60s. Well, tonight *20/20* _________ into the weird world of Andy Warhol.

ALL ABOUT ANDY WARHOL

As you watch the segment, take notes to answer the questions below: 01:36:25-01:43:20

1. When Andy Warhol says that he thinks it would be great to support Jimmy Carter and Teddy Kennedy, he give support to Downs's statement that ______________________________

   ____

2. Wherever he goes, Warhol carries ______________________________
3. Warhol's attitude toward celebrities is ______________________________

   ____

4. Concerning Andy Warhol and fame, the writer Fran Liebowitz says ____

   ____

5. Before he was known as an artist, Andy Warhol was ______________

   ____

6. Andy Warhol's experimental films sometimes lasted ______________
7. Bored, outraged audiences sometimes demanded ______________
8. In 1968, an actress who appeared in one of Warhol's films __________

   ____

9. For doing a commissioned portrait, Andy Warhol is paid __________
10. Andy Warhol's book *Exposures* shows ______________________________

    ____

11. Concerning Andy Warhol, the performer Liza Minnelli says ________

    ____

12. The Whitney Museum in New York ______________________________

ANDY WARHOL'S EXPRESSIONS

As you watch this part of the segment with the sound off, describe the changing activities and also the expressions on Andy Warhol's face as he travels to the very diverse parts of the United States on his book-signing tour. What do you suppose is the reason for his varied facial expressions? Work with a partner and write a description below. 01:41:31-01:42:41

____

____

HOW DID HE DO IT?

01:44:23-
01:47:20

In this part of the segment, the narrator describes the process Andy Warhol used to create his portrait of Farrah Fawcett. Describe this process. Use transitional words such as *then*, *after that*, *next*, and *finally* in your description. The first sentence has been done for you.

First, Andy Warhol took photographs of his subject with a Polaroid camera.

Next, ______________________________

____

After that, ______________________________

____

Finally, ______________________________

____

THE THREE ANDY WARHOLS

01:47:20-
01:47:39

In this part of the segment, Hugh Downs talks about the three Andy Warhols, labeling him as a different type of artist during each period of his career. First, write the three labels given to Andy Warhol in the space below. Then read the three descriptions that follow and match the description with the correct label.

THE THREE ANDY WARHOLS

______ 1. ______________________

______ 2. ______________________

______ 3. ______________________

a. painted dollar bills, Coca-Cola bottles, Brillo boxes, and Campbell's soup cans.
b. painted portraits of Mick Jagger, Mohammed Ali, Liza Minnelli, and Farrah Fawcett.
c. drew shoes for magazines like *Glamour*, *Vogue*, and *Harper's Bazaar*.

Language Focus

VOCABULARY IN CONTEXT

Guess the meanings of the *italicized* words below. First circle the words in the sentence that helped you to guess the meaning. Then write a definition in the space provided.

1. **Downs:** Quiet and *inarticulate,* Andy is content to be an observer.
   Warhol: I look shy, and I am shy. And my voice is cracking, shy.

   inarticulate: ____

2. Andy has always had a *fan mentality:* he worships stars, gets to know them.

   fan mentality: ____

3. With his distinctive, colorless face, Andy always *stands out.*

   stands out: ____

4. He's really *elevated* fame to a higher level. Fame is more famous now than it was before there was Andy.

   elevated: ____

5. After ten years of losing money, the magazine is finally *turning a profit.*

   turning a profit: ____

DESCRIBING AND GUESSING: USING THE READ AND LOOK UP METHOD

One part of the segment shows the many different celebrities photographed and painted by Andy Warhol. Choose from the list on page 116 and write a description of one of these celebrities. Then read your description aloud and the rest of the class can guess which of the celebrities is being described. When reading your description, use the "read and look up" method. Look at the example below.

> Who is one of / the best known performers / in America today? I'll give some hints. / She's danced in/Broadway shows, / sung in nightclubs and appeared in movies. / She is famous / for her role / in a film / called, *Cabaret.* / She has / short, dark hair / and enormous eyes. / She is the daughter / of a famous American actress /and singer / named Judy Garland. /

Did you guess the name of the performer described in the above passage? Her name is checked in the list of celebrities on page 116.

Jimmy Carter	Marilyn Monroe
Farrah Fawcett	Elizabeth Taylor
Mohammed Ali	✔ Liza Minnelli
Mick Jagger	

Now write a description of one of the other celebrities in the list above. If you do not know the celebrities, pick another well-known personality.

____

____

____

____

Postviewing

DO YOU AGREE OR DISAGREE?

Work with a small group of students. Discuss whether you agree or disagree with the following statements and why. Choose one student to be the group leader. The leader can report the opinion(s) of the group to the rest of the class.

1. I like Andy Warhol's paintings.
2. Andy Warhol's assembly-line portraits and pop culture subject matter make an important statement about life in modern society.
3. Andy Warhol made important contributions to the world of art.
4. Andy Warhol probably influenced artists in my country.
5. Andy Warhol was not a real artist.
6. Andy Warhol's best art was publicity.

RELATED READING 1

Read the article *The Death of Andy Warhol* to answer the questions that follow.

The Death of Andy Warhol

On the morning of February 23, 1987, the artist Andy Warhol died at the age of 49 at New York Hospital in New York City. Warhol had been admitted to the hospital the day before for gall bladder surgery. Although doctors had expected this to be a simple, routine operation, Warhol died just hours later from complications related to the surgery.

Andy Warhol left behind a major twentieth-century artistic legacy. Whether or not one appreciates Warhol's work, few would deny its impact on the art of recent decades, or the fascination it aroused in viewers. After Warhol's death, the artist's collection of works sold for $25.3 million at the world-famous Sotheby's art auction house in New York City. When a journalist asked the writer Fran Liebowitz, a friend of the artist, to guess what Warhol's reaction would have been to the sale of his work for such an enormous sum of money, Liebowitz replied, "Andy must be so furious that he is dead."

Source: Bockris, Victor. *The Life and Death of Andy Warhol.* Bantam Books, 1989.

QUESTIONS

Scan the text and underline those parts that contain answers to the questions below. Write the number of the corresponding question next to the underlined text. Then, using your own words, write short answers to the questions below. Next, ask your partner to answer the questions orally, without looking at the text or written answers.

1. When, where, and at what age did Andy Warhol die?

____

____

2. How did he die?

____

____

3. What kind of legacy did Andy Warhol leave behind?

____

____

4. How much did Warhol's collection of works sell for at Sotheby's auction house?

____

____

5. What reaction did Fran Liebowitz imagine her friend Andy Warhol would have had to the sale of his work?

____

____

RELATED READING 2

Read the article *Warhol's Works Live On* to prepare for the task that follows.

Warhol's Works Live On

Andy Warhol once said, "It will take my death for the Museum of Modern Art to recognize my work." Andy's statement turned out to be true. It was not until two years after his death in February 1989 that New York City's world-famous Museum of Modern Art (or MOMA, as it is abbreviated) held a retrospective of Andy Warhol's work. MOMA has some of Warhol's most famous pieces in its permanent collection: among them, *Golden Marilyn Monroe, Self-Portrait*, and two of his famous Campbell's soup paintings.

In 1971, New York's Whitney Museum of American Art had also held a retrospective of the artist's work. The Whitney is still the home of some of Warhol's most important pieces of work, among them *Green Coca-Cola Bottle, Ginger Rogers,* and *Bow-Ties*.

QUESTIONS

Write your own questions to accompany the above text. Begin your questions with the words below. After writing your questions, ask your partner to answer them without looking at the text.

1. What ______________________________________?
2. When ______________________________________?
3. Which ______________________________________?
4. Where ______________________________________?

DIFFERING VIEWS OF ANDY WARHOL

In the segment we hear opposing views of Andy Warhol's work by two art historians, John Richardson and William Walton.

John Richardson's view of Warhol's work is positive:

> "Andy was the first pop artist to use cans of soup, Brillo boxes, bottles of Coca-Cola, I mean all kinds of things that ordinary people don't think of as being worthy of artistic treatment. And I think that Andy took these and somehow he made them into unforgettable images. He gave them a completely new look."

William Walton's view of Warhol's work is much less positive:

> "I don't consider it a feast of great painting. He hasn't contributed much to the development of interesting things in American painting. One or two [of his paintings] seem amusing and original, but a gallery full is, finally, a big bore."

Form small groups and discuss which of the above views you are inclined to agree with. ("inclined to agree" means that, although you have not made an absolute judgment on an issue, your feeling goes in a particular direction). Justify your point of view by referring to information you have learned in the video segment. Complete the statements below to talk about your opinion.

I'm inclined to agree with Richardson/Walton because . . .

OR

I'm inclined to disagree with Richardson/Walton because . . .

GROUP PROJECT: MAKE A RADIO PROGRAM

Work with a small group of students and prepare a script for a radio program. Choose one of the quotations below as the theme of your program. In your program, discuss the significance of the statement. What do you think Andy Warhol meant by it? Do you agree or disagree with what he said? Why? Radio programs can be from 5 to 15 minutes in length. When the programs are finished, have a class contest to judge the best program.

> "There will come a time in America when everyone is famous for fifteen minutes."

> "I like to make people look good; I think that's what life is all about."

When asked to defend himself against his critics, Andy Warhol replied:

> "Oh, I can't. They're right."

WRITING AND RESEARCH PROJECTS

1. Choose one of the portraits or other works by Andy Warhol shown in the video segment and write a paragraph describing it.
2. Think about some elements of popular culture in your own society. Could any of them capture something important about life in contemporary times and be subjects for paintings? Prepare a written or oral report explaining why you feel these could serve as icons of modern times. Be sure to use some of the new vocabulary words you have studied in this segment.

SUGGESTED READINGS

Warhol Andy, *Exposures*. Grosset & Dunlap, 1979.

Ratcliffe, Carter. *Andy Warhol*. Abbeville Press, Modern Masters Series, 1983.

Gidal, Peter. *Andy Warhol: The Factory Years*. DaCapo, 1971.

Warhol, Andy. *The Philosophy of Andy Warhol: From A to B and Back Again*. Harcourt Brace Jovanovich, 1975.

Segment 11

Martha Graham

from *World News Tonight,* 5/11/90
Runtime: 4:57
Begin: 01:48:07

Martha Graham was perhaps the greatest U.S. pioneer in the development of modern dance.

Previewing

KEY QUESTIONS

1. What motivates an artist to create a new method of expressing his or her art?
2. How does an artist risk his or her career when presenting a new or innovative idea?
3. In defining success, how much importance should be placed on public acceptance of an artist's work?

DISCUSSION QUESTIONS

1. Do you like dance? Do you enjoy dancing yourself or watching performers dance? What kind of dance do you like?

    ____

    ____

2. What is your opinion of modern dance versus traditional dance?

    ____

    ____

3. Some dancers are controversial because they don't follow conventional ideas about dance. Can you think of any controversial dancers?

___

___

PREDICTION

Based on the title of the segment, *Martha Graham*, the discussion you've had with your classmates, and your own knowledge, what information do you think will be included in the video segment?

1. ___
2. ___
3. ___
4. ___

ESSENTIAL WORDS TO KNOW

The *italicized* words below are used in the video segment. Read the sentences. Then match the words to their meanings on page 123.

disciplinarian:	Linda's father was a strict *disciplinarian* and told her that she always had to obey him.
choreographer:	The *choreographer* made up a dance for the ballet.
raved:	The movie was so good that people *raved* about it.
booed:	The audience didn't like the modern dance routine and loudly *booed* the dancers off the stage.
vegetated:	Because he was depressed and inactive, he sat by the television set all day and *vegetated*; he didn't use his energy in a productive way.

______ 1. disciplinarian — a. to give high praise

______ 2. choreographer — b. to be inactive, with little physical or mental activity

______ 3. rave — c. a person who believes in enforcing strict rules

______ 4. boo — d. a person who designs and plans a dance

______ 5. vegetate — e. to make a sound that expresses disapproval

Global Viewing

CONFIRMING YOUR PREDICTIONS

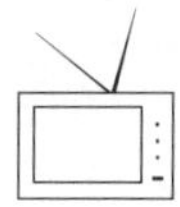

01:48:14-
01:52:56

Watch the entire video segment. As you watch, put a check (✓) next to any of your predictions from the preview section that were mentioned.

Prediction 1 ______

Prediction 2 ______

Prediction 3 ______

Prediction 4 ______

MAIN IDEA

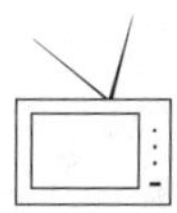

01:48:14-
01:52:56

Imagine that some of your classmates have missed this program. If you could tell them the main idea of the program in just one or two sentences, what would you say? Write it below. Then compare your sentence with that of other students in the class.

____

____

THINK, WRITE, AND DISCUSS

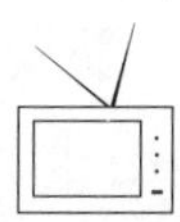

01:48:14-
01:52:56

Watch the entire segment and answer the questions below. If you wish, you may take notes on a separate sheet of paper. After you've finished, discuss your answers with the class.

1. What is your impression of Martha Graham? What adjectives would you use to describe her?

   ____

   ____

2. Did Martha Graham always want to be a dancer? Did she have the support of her family?

____

____

3. How were Martha Graham's ideas about dance different from conventional ideas? What did she dislike about the ballet?

____

____

4. Was Martha Graham always successful with audiences and critics? Where did she perform?

____

____

5. What was Martha Graham's attitude toward retirement?

____

____

Intensive Viewing

MARTHA GRAHAM'S LIFE

01:48:30-01:52:53

As you listen to this part of the segment, take notes to supply the information below.

1. How Martha Graham describes
   a. modern dance: ______________________________
   ____
   b. performance: ______________________________
2. Martha Graham's childhood
   a. Where she was born: ______________________
   b. Where she was raised: ______________________
   c. What her father was like: ______________________

3. Martha Graham's early dance career
    a. Where she studied: ________________________________
    b. When she studied: ____________________
    c. How her teacher described her: ____________________________
4. The Martha Graham Dance Company
    a. When it was formed: __________________
    b. What it did for American dance: __________________________
    c. What form and movement should express:
        1. ___________________________
        2. ___________________________
5. Martha Graham during the 1940s, 1950s and 1960s
    a. How she spent her time: ________________________________
    ____
    b. How one audience reacted in Italy in the '40s: ________________
    ____
6. What Agnes de Mille said about Martha Graham: ________________
    ____

Language Focus

VOCABULARY CLOZE

Write the appropriate words in the blank spaces in the dialogue below and on page 126. Be sure to use the correct form of the word you choose. Then watch this part of the segment and check your answers.

01:48:14-
01:49:12

be	have	catch	heritage
define	perfect	secure	treasure

Peter Jennings: Finally, on May 11, 1990, our Person of the Week. It was her birthday which ___________ our attention. She _____ 96 years old today, and there is simply no question that she has ____________ her place in history, for she _______ created a permanent part of a national cultural _______________. She is a dancer.

Martha Graham: I only know this: you dance because you have to.

Peter Jennings: Martha Graham is to dance as Picasso was to art. During nearly eight decades in the theater she has _______________ and ____________________ the art form that is modern American dance.

Martha Graham: I'd describe it as an attitude toward the body, loving the body, exhibiting it, and ______________________ it.

STRUCTURE FOCUS: PRESENT PERFECT/PAST PERFECT

The present perfect (*has* or *have* and the past participle of the main verb) describes an activity that took place in the past and continues in the present. Look at the example below.

Nancy *has lived* in Paris for ten years.

In this example, Nancy still lives in Paris.

The past perfect (*had* and the past participle of the main verb) describes an activity that began in the past and ended at a later point in the past. This activity also began and ended before another point in the past. Look at the example below.

Nancy *had lived* in Paris for ten years before she moved here.

In this example, Nancy no longer lives in Paris. She lived there before moving here, which also happened in the past.

In the segment, Peter Jennings begins by telling the viewers that it is Martha Graham's birthday today. He then describes the various things she accomplished up to that point in her life by using the present perfect tense.

There is simply no question that Graham *has secured* her place in history because she *has created* a permanent part of the national cultural heritage.

In April 1991, a year after the program was recorded, Martha Graham died. If Peter Jennings were describing the accomplishments of her career now, while referring to her birthday (past) in the segment, he would use the past perfect tense.

There is simply no question that Graham *had secured* her place in history because she *had created* a permanent part of the national cultural heritage.

Notice that it is not necessary to refer to the segment airdate (past).

Below are some of the achievements of Martha Graham's career. Make complete sentences using the past perfect. Your past reference point is the segment airdate. The first one has been done for you.

1. revolutionized American dance

   In the course of her career, Martha Graham had revolutionized American dance.

2. taught dancers to express more emotion and drama in dance

   ____

   ____

3. was a "towering presence" and an inspiration to many young dancers

   ____

   ____

4. studied with the famous Denishawn Dance Company

   ____

   ____

5. danced with the Greenwich Village Follies

   ____

   ____

6. created the Martha Graham Dance Company

   ____

   ____

Now work with a partner and write one or two sentences describing activities and accomplishments in Graham's career based on information you learned in the program.

1. ______________________________

2. ______________________________

GENERATING DESCRIPTIONS

01:48:38-01:52:14

Watch the program again with the sound and closed captions turned off. Pay careful attention to the dance performances. Write as many adjectives as you can to describe how the dances look and how watching them makes you feel. (You may use a dictionary if you need to.)

Work with a partner and compare your list of adjectives. Are your choices of words the same or different? Together, choose the adjectives you feel are the most appropriate. Write a poetic paragraph in the space below describing Martha Graham's dances and how they make you feel. Then read the paragraphs aloud to the class. If you wish, you may practice using the read-and-look-up method.

Which pair of students has written the most poetic paragraph? Choose several of the more poetic paragraphs and use them to narrate one of the dance performances in the segment. Turn off both the sound and closed-captioning and watch as students read their original program narratives.

Postviewing

DO YOU AGREE OR DISAGREE?

Work with a small group of students. Discuss whether you agree or disagree with the following statements and why. If there is a difference of opinion among the members of your group, justify your answer to the other group members. Choose one student to be the group leader. After the small group discussion, the leader can report the opinions of each member of the group to the rest of the class.

1. Martha Graham seems like an ordinary person.
2. Great artists always have eccentric personalities.
3. Martha Graham reminds me of certain innovative artists in my country.
4. Modern art is more interesting than traditional art.
5. Martha Graham's dances don't look so modern in this day and age.

RELATED READING 1

Read the article *Ruth St. Denis: American Dance Pioneer* and answer the questions that follow.

Ruth St. Denis: American Dance Pioneer

Ruth St. Denis, born in New Jersey in 1877, is known as one of the pioneers of modern American dance. Along with her husband, Ted Shawn, who was also an innovative dancer and choreographer, St. Denis helped American dance find its own style, independent of the European dance tradition that Americans had long imitated. It is interesting that St. Denis's inspiration with the idea for her unique dances came from — of all things — an advertisement poster for cigarettes! The poster showed a picture of the Egyptian goddess Isis. The picture of the goddess fired St. Denis's imagination, and she decided to be a concert artist specializing in Oriental dances.

St. Denis and Shawn, with their Denishawn Dance Company, first introduced American audiences to the dance aesthetics of the Orient through their exotic performances. These dances showed a combination of Eastern and Western sensibilities. The company danced barefoot on stage in Oriental costumes and made much use of scarves, veils, dramatic poses, and flowing arm movements. The Denishawn Dance Company toured the United States and Europe and performed to enthusiastic audiences.

Because St. Denis and Shawn believed that dancers should be trained in the dance traditions of Eastern as well as Western cultures, they started the Denishawn School in 1915. The most renowned pupil to graduate from the Denishawn School was Martha Graham.

Source: *Encyclopedia Americana*, Vol. 8. Grolier, Inc., 1989.

QUESTIONS

Scan the reading passage to locate answers to the questions below. When you have located the parts of the text containing the answers, underline them and write the number of the corresponding question next to the underlined parts. Then write answers to the questions below.

1. Who was Ruth St. Denis?

____

____

2. Where was she from?

____

____

3. What first inspired Ruth St. Denis to make dances?

____

____

4. What was unusual about the Denishawn Company's dances?

____

____

5. When did the Denishawn School open?

____

____

RELATED READING 2

Read the article *Isamu Noguchi: Artist of the Twentieth Century* to prepare for the task on page 131.

Isamu Noguchi: Artist of the Twentieth Century

Isamu Noguchi, one of the world's great sculptors and designers, was born on November 17, 1904, in Los Angeles, California. Noguchi, a Japanese-American, spent his early years in Japan but later returned to the United States and studied art in New York. He was inspired by the moveable abstract sculptures of the great American sculptor Alexander Calder. Like Calder, Noguchi showed the expressive power of abstract shapes in his work. Noguchi was also influenced by the abstract works of the sculptor Giacometti and the painters Picasso and Miro whom he met during his years working in Paris, where he designed the garden for the United Nations Educational, Scientific, and Cultural Organization (UNESCO). The emotional force of Noguchi's moveable abstract forms seemed a perfect match for Martha Graham's powerful, concentrated choreography. Noguchi designed many of the stage sets and costumes for Graham's dances, including the famous wire dress for her powerful ballet *Cave of the Heart*.

Noguchi's works for the stage and his works of sculpture may be seen at museums around the world, including the Metropolitan Museum of Art and the Museum of Modern Art in New York. New York City is also the location of the Isamu Noguchi Museum, where many of the artist's greatest sculptural works and sculpted gardens can be viewed.

Source: *Encyclopedia Americana*, Vol. 20. Grolier, Inc., 1989.

QUESTIONS

Write questions about the reading on page 130. Begin each of your questions with the words below. Then ask your partner to answer the questions without looking at the text.

1. Who ____?

2. When ____?

3. Where ____?

4. What ____?

5. Why ____?

NOTABLE QUOTES

Write your reaction to one of the following quotations.

1. At the end of the program, when asked if she had any plans to retire, Martha Graham answered,
   "To retire to what? What do you do? Vegetate or reminisce or something? No, I want to exist now, in the moment."
   Do you think most people would feel the same way as Martha Graham did about retiring from their work? How do you imagine you will feel when it is time for you to retire? Will you want to reminisce? Or will you want to, as Graham put it, "exist in the moment"?

2. In her autobiography, *Blood Memory*, Martha Graham says,
   "I chose not to have children for the simple reason that I felt I could never give a child the caring upbringing which I had had as a child. I couldn't control being a dancer. I knew I had to choose between a child and dance, and I chose dance."
   Martha Graham felt she had to choose between devoting herself to a career and devoting herself to having a child. She did not feel it was possible to do both. Do you agree that in order for people to make major accomplishments, their energy cannot be divided between pursuing their work and caring for their families? Do women always have to make a choice between a career and a family? Do men have to make this choice too? Why or why not?

PROJECTS

IDEA #1 – COMPARE CLASSICAL BALLET TO MODERN DANCE

Go to your local library and borrow one or more videos of classical ballet. Watch these videos with the entire class. As you are watching, keep in mind how Martha Graham felt. "She considered the classical ballet of the day little more than sleepwalking." What do you think of classical ballet? How does it compare to some of the dances you saw in this segment? Do you like it more or less?

Now rate the video. Give the video either a thumbs-up (meaning you liked it) or a thumbs-down (meaning you did not like it). Be prepared to defend your rating.

IDEA #2 – WRITING AND RESEARCH PROJECTS

Find out more about the history of modern dance by researching one of the items below.

1. The Denishawn Dance Company School
2. The Martha Graham Dance Company
3. Isamu Noguchi
4. Twyla Tharp

SUGGESTED READINGS

Dore, Ashton. *Noguchi: East and West*. Knopf, 1992.

Graham, Martha. *Blood Memory: An Autobiography*. Doubleday, 1991.

De Mille, Agnes. *Martha: The Life and Work of Martha Graham*. Random House, 1991.

Saint-Denis, Ruth. *An Unfinished Life: An Autobiography*. Harper, 1939.

Segment 12

Black Filmmakers

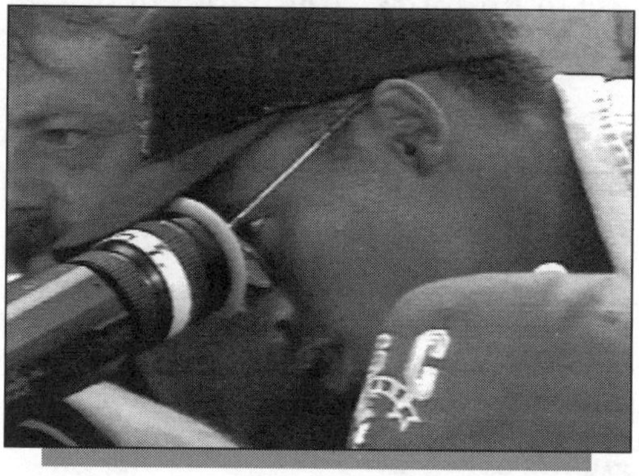

from *Nightline,* 6/21/91
Runtime: 7:48
Begin: 01:53:12

The black experience in America has become a profitable movie market, and the Hollywood establishment is taking notice.

Previewing

KEY QUESTIONS

1. How do stereotypes hurt the people they are directed against?
2. How do stereotypes hurt the people who believe in them?
3. What can a society or government do to discourage stereotyping and discrimination?

DISCUSSION QUESTIONS

1. Can you name some famous American films? Can you also name the directors of these films?

    ____

    ____

2. How many of the film directors you named are white men? Can you think of any films directed by members of minority groups (i.e. African American, Asian American, Native American, women)?

    ____

    ____

3. Why do you suppose there have been so few American film directors from minority groups?

____

____

4. How has the change in social standing of minority groups in America affected the number of people from these groups who become film directors?

____

____

PREDICTION

Based on the title of the segment, *Black Filmmakers*, the discussion you've had with your classmates, and your own knowledge, what information do you think will be included in the video segment?

1. _______________
2. _______________
3. _______________
4. _______________

ESSENTIAL WORDS TO KNOW

The *italicized* words will be used in the video. Read the words as they are used in the following sentences and try to come up with your own definition.

1. People sometimes say things that are *demeaning* and insulting to others without realizing it.

   demeaning: _______________

2. Until the American Civil War, African Americans had to suffer the *indignity* of slavery.

   indignity: _______________

3. One harmful *caricature* of African American men in films of the 1930s commonly showed them as good natured but not very intelligent.

   caricature: _______________

4. Pacifists often believe it's best to *turn the other cheek* rather than respond violently or angrily.

   turn the other cheek: ____________________

5. People sometimes think of members of other races as parts of *monolithic* groups rather than as unique individuals.

   monolithic: ____________________

Global Viewing

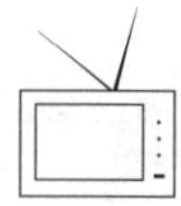

01:53:12-
02:01:00

CONFIRMING YOUR PREDICTIONS

Watch the entire video segment. As you watch, put a check (✓) next to any of your predictions that were mentioned.

Prediction 1 ____

Prediction 2 ____

Prediction 3 ____

Prediction 4 ____

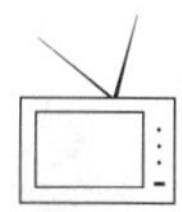

01:53:12-
02:01:00

MAJOR POINTS/ MAIN IDEA

Write down a few of the major points made in the program.

Imagine that a classmate of yours has missed this program. He asks you to give him a short summary of it. What would you tell him in one or two sentences? Write your summary below.

THINK, WRITE, AND DISCUSS

01:53:12-
02:01:00

Watch the entire segment and answer the questions below. If you wish, you may take notes on a separate sheet of paper. After you've finished, discuss your answers with the class.

1. How has the face of black film changed in the last 20 years?

   ____

   ____

2. How did Hollywood portray blacks on film in the '30s, '40s, '50s, and '60s?

   ____

   ____

3. What opportunities and resources are now available for blacks in film that were not available in the past?

   ____

   ____

Intensive Viewing

TRUE OR FALSE

01:54:07-
02:00:46

Work with another student and tell if the following statements are true or false. Be prepared to justify your answers by referring to specific parts of the program

____ 1. *Amos 'n Andy* was a harmless, good-natured program that both blacks and whites enjoyed.

____ 2. Recent films by young black filmmakers have cost a lot of money to make and have not made very large profits.

____ 3. Fifty years ago, black films were made for black audiences.

____ 4. Thirty years ago, black actors could portray the hero of a film, but the heroes were defined by white standards.

____ 5. Director Melvin Van Peebles was looking for a job as an elevator operator when he first came to a Hollywood film studio in 1938.

____ 6. Filmmaker Matty Rich says the suffering he saw in his community inspired him to make films.

____ 7. Filmmaker John Singleton says that he feels a white film director could also have made a movie like his *Boyz 'n the Hood*.

____ 8. The movie industry is likely to judge films made by young black filmmakers by the same standard they use to judge films by white directors.

SOUND OFF

01:55:19-01:56:10

In this part of the video you will see three short film clips showing African Americans in American films. With the sound turned off, watch the clips, taking notes about how African Americans are portrayed. Then write a comparison of the three clips, summarizing how the portrayals have changed over time.

First film (*Song of Freedom*, 1930s): ______________________________

____

Second film (*Shaft*, 1970s): ______________________________

____

Third film (*Jungle Fever*, 1990s): ______________________________

____

Summary: ______________________________________

____

____

FILMS ABOUT AFRICAN-AMERICAN LIFE

01:56:10-01:58:55

Watch this part of the segment and match the title of each film with the correct description.

a. *A Rage in Harlem*
b. *The Five Heartbeats*
c. *Straight Out of Brooklyn*
d. *New Jack City*
e. *Guess Who's Coming to Dinner?*

_____ 1. A coming-of-age story made on an investment of $70,000 supplied by family, friends, and associates.

_____ 2. A romantic comedy thriller.

_____ 3. A portrait of urban drug wars, made by the film company Warner Brothers, for $10,000,000.

_____ 4. A bittersweet tale of a 1960s singing group.

_____ 5. A film featuring a black hero — but defined by white people's standards.

IDENTIFY THE FILMMAKER

Watch this part of the segment and match the name of each filmmaker with the correct description.

01:56:10-01:58:55

a. Matty Rich
b. John Singleton
c. Spike Lee
d. Mario Van Peebles
e. Robert Townsend
f. Bill Duke

_____ 1. Twenty-three-year-old director of *Boyz 'n the Hood*
_____ 2. Forty-eight-year-old director of *A Rage in Harlem*
_____ 3. Nineteen-year-old director of *Straight Out of Brooklyn*
_____ 4. Director of the box office successes *Jungle Fever* and *Do the Right Thing.*
_____ 5. Thirty-four-year-old director of *The Five Heartbeats*
_____ 6. Thirty-one-year-old director of *New Jack City*

WHO SAID IT?

With a partner, match the names of the people from the list with the sentences that best restate their ideas. If you need to, watch the segment again.

01:58:25-01:59:33

a. Matty Rich
b. Jeff Greenfield
c. Mario Van Peebles
d. John Singleton

_____ 1. We came into a black neighborhood with a black crew and a black cast; it was like a homecoming.
_____ 2. The Hollywood establishment thinks we're all on the same basketball team. They say, "Black folks are making money. Let's throw some dollars out there."
_____ 3. I made this movie because when I looked outside my window, I saw pain.
_____ 4. Hollywood's favorite color has always been green.

Language Focus

FILM VOCABULARY

Certain words and expressions are often used when talking about or writing about film. Some of those words and expressions are used in the program on black filmmakers. Work with a classmate and identify the meanings of the words and phrases below:

1. a box office smash ____________________
2. a coming-of-age story ____________________

3. an intensely personal vision ______________________________

4. audience ______________________________

5. cast ______________________________

6. crew ______________________________

WORD FORMS

The following words were used in the segment. Complete the chart below by adding the other forms of the given words.

NOUN	VERB	ADJECTIVE
1. face	__________	__________
2. __________	__________	romantic
3. stereotype	__________	__________
4. __________	(none)	comic
5. __________	(none)	racial
6. __________	__________	assertive
7. vision	__________	__________
8. __________	to emphasize	__________

Postviewing

DO YOU AGREE OR DISAGREE?

With the members of your group, discuss whether you agree or disagree with the following statements. Give reasons to justify your opinions.

1. Only black filmmakers can make good films about the black experience.
2. Filmmakers and other artists should make art based only on their own personal experience.
3. I have seen one or more films made by African American filmmakers.
4. Members of minority groups are often portrayed by negative stereotypes in films and other art forms.
5. The experience of social, economic, or political oppression often gives birth to fine works of art.

RELATED READING

Read the following article to answer the questions that follow.

Spike Lee: A King Among Young Filmmakers

Of all the past decade's films by young, talented black filmmakers, those of director Spike Lee that have received the most attention. Born in 1957, Lee first came to national attention -with his low-budget, black-and-white surprise box office hit, *She's Gotta Have It*, about an independent young black woman and her three rival suitors. Among the films that followed were the popular, if controversial, *Do the Right Thing* which explores the making of a race riot in an inner-city neighborhood of New York, and *Jungle Fever,* which focuses on the romantic relationship between a black man and a white woman. Lee's movies are not only thought-provoking but very entertaining, and thus draw U.S. audiences of all races and backgrounds. As Lee said in an interview, his films have put "race relations in this country back on the agenda . . . because there are far too many people walking around that think that racism is a dinosaur that's a thing of the past."

QUESTIONS

Scan the reading passage to locate the answers to the questions below. When you have located the parts of the text containing the answers, underline them and write the number of the corresponding question next to the underlined parts. Then write answers to the questions below.

1. Who is Spike Lee?

___

___

2. When was Spike Lee born?

___

___

3. Which of his films first brought him to national attention in the U.S.?

___

___

4. What was Lee's most controversial film?

___

___

5. What does Lee say about his films?

___

___

STEREOTYPES

What Are Stereotypes?

A stereotype is an overly simplified view of another person — a view based on generalized, pre-conceived ideas about that person's racial or cultural background. Every racial or cultural group has stereotypes of other racial and cultural groups. Stereotypes are an attempt to simplify what in reality is one of the most complex phenomena in the world — other human beings and their cultures. When we stereotype other people, we do not see them as individuals; rather, we see them only in terms of the pre-conceived ideas that our own cultural or racial group has about them.

In the program segment, Donald Bogle talks about how black people have often been stereotyped in American movies. He mentions three major film stereotypes of black people. Match the stereotypes in column A with their correct definitions in column B.

COLUMN A	COLUMN B
comic coons	Black men who showed a submissive attitude toward white people, behaving as if white people were their superiors
[Uncle] Toms*	Black people who behaved in a silly, slow-witted manner, which made them appear ridiculous and funny to white audiences
feisty mammies	Black women servants who were very fat, whose personalities were rather high-spirited and argumentative, and who were aggressive (although these women always had a lot of genuine love for their white superiors)

*The term, "Uncle Tom" is originally from the title of a famous nineteenth-century novel by Harriet Beecher Stowe named *Uncle Tom's Cabin*. The novel focuses on the story of Uncle Tom, a slave who is badly mistreated by his white owners, for whom he shows only love and loyalty. (The term Uncle Tom has had a derogatory meaning for the last three decades or so; however, perhaps ironically, the novel from which the term derived had a great influence on promoting the anti-slavery movement in the United States.)

Think of some of the ways in which your culture stereotypes members of other cultural, racial, or national groups. Below are some typical stereotypes of Americans.

Americans don't care about their families.

American wives always dominate their husbands.

What stereotypes do you think people from other cultures have about people in your culture?

NOTABLE QUOTES

Prepare an oral or written report, incorporating some of the new words you learned in this program. Look at the quotations below and the questions that follow to prepare for your report.

"I don't want to be known as a black woman filmmaker. I want to be known as what I do — not by what's self-evident on the outside."

— Stephany Minor, New York University Film School Graduate

"Race has been a dominant part of history from Moses until now, so I don't think it's ever going to change. It's always going to be a part of the vocabulary for film or any other business."

— Darryl McCane, Columbia University Graduate Student

1. With which of the above "notable quotes" do you tend to agree? Should filmmakers or other artists always be viewed in terms of their racial and cultural backgrounds? Or should artists' work be judged apart from such personal characteristics?
2. Is it possible for a member of one racial or cultural group to write convincingly about members of a different racial or cultural group? For example, is it possible for a white person to write realistically about black people? Is it possible for Westerners write stories about Asian people or vice versa?

PROJECT

Write about a minority group in your country and explain the relationship this group has to the larger, mainstream society. Has this group been stereotyped in any way by the larger culture? In what way? Do you think the members of this minority group and those of the mainstream society understand each other well? Why or why not? Describe particular features of the group's culture with which you are familiar.

SUGGESTED READINGS

Bogle, Donald. *Toms, Coons, Mulattoes, Mammies, & Bucks: An Interpretive History of Blacks in American Films*. Viking, 1973.

Blacks in American Film & Television: An Illustrated Encyclopedia. Simon & Schuster, 1989.

Fredrickson, George. *Black Image in the White Mind: The Debate on Afro-American Character and Destiny, 1817–1914*. Wesleyan University Press, 1987.

Gaines, Edith M., et al. *Black Image Makers*, New Day Press, 1988.